CONCRETE
PROBLEM CLINIC

The Aberdeen Group®
a division of Hanley-Wood, Inc.

426 S. Westgate St., Addison, Illinois 60101
Telephone: 630-543-0870 Fax: 630.543.3112

Concrete Problem Clinic

Published by The Aberdeen Group
426 S. Westgate St.
Addison, IL 60101

The Aberdeen Group and its employees and agents are not engaged in the business of providing architectural or construction services, nor are they licensed to do so. The information in this book is intended for the use of engineers and contractors competent to evaluate its applicability to their situation, and who will accept the responsibility for the application of the information. The Aberdeen Group and the author disclaim any and all responsibility for the application of the information.

Managing Editor: Patricia L. Reband
Art Director: Joan E. Moran
Cover photography: Bruce A. Suprenant; Structural Preservation Systems Inc.
Publisher: Mark DiCicco

ISBN: 0-924659-24-6
Item No. 4061

Contents

How to order additional copies

Additional copies of *Concrete Problem Clinic*, *Masonry Problem Clinic* and the previously published *Concrete and Masonry Problem Clinic* are available through The Aberdeen Group. Call toll free 800-323-3550. Fax your order 24 hours a day: 630.543.3112. You may also order electronically at www.tagbookstore.com. Visa, MasterCard and American Express are accepted.

Introduction

The Problem Clinic in *Concrete Construction*®, Repair Q & A in *Concrete Repair Digest*®, Production Troubleshooting in *Concrete Journal*® and Troubleshooting in *The Concrete Producer*™ answer readers' questions about the construction, maintenance and repair of concrete. Experienced editors, technical experts and literature in the field all contribute to the answers and solutions.

This collection includes answers from *Concrete Construction*® from 1989 to 1998, *Concrete Repair Digest*® from 1990 to 1998, *Concrete Journal*® from 1990 to 1996 and *The Concrete Producer*™ from 1997 to 1998. Information is grouped in the following categories:

- Admixtures
- Aggregate Exposure—Sandblasting/Waterblasting
- Colored Materials
- Concrete Joints
- Crack Repair
- Curing
- Defects—Horizontal Surfaces
- Defects—Vertical Surfaces
- Durability
- Finishing
- Formwork
- General Repair
- Materials
- Mixing
- Overlays
- Paving
- Placing
- Production
- Reinforcement
- Safety
- Stain Removal/Cleaning
- Testing
- Tolerances
- Miscellaneous

Where appropriate, the original response is altered to reflect dates for the latest documents from the information sources cited.

Acknowledgments

We would like to thank the following people for their help in producing *Concrete Problem Clinic*.

Ward R. Malisch, Editorial Director and Engineering Editor, *Concrete Construction*® magazine

Bruce A. Suprenant, Vice President of Research and Development for The Aberdeen Group

Rick Yelton, Senior Engineering Editor, *The Concrete Producer*™ magazine

Timothy S. Fisher, formerly Engineering Editor for *Concrete Construction*® and *Concrete Journal*® magazines

Jeff Ende, Associate Engineering Editor, *Concrete Construction*® magazine

Chapter 1

Admixtures

Q. *Our standard flatwork mix is six-bag, ¾-inch limestone, with 6% air and 1½ pounds of polypropylene fibers. It is placed at a slump of 4 inches. When ambient temperatures dictate, we use a maximum of 1% calcium chloride. Adding the chloride at the ready mix plant ensures adequate mixing, but results in the need to add water at the jobsite to maintain slump, something we are reluctant to do. Since adding chloride flakes to the mixer at the site carries its own potential problems, we premix the flakes in a bucket, then pour the solution into the mixer. Mixing the flakes in a bucket beforehand creates tremendous heat; energy that seems to be lost before the flakes are mixed with the concrete. Should we adjust the quantity of the chloride flakes to compensate for this, and does our calcium chloride solution have a shelf life before it is mixed into the concrete?*

A. We asked Guy Detwiler of Chicago's Material Service Corp. to respond to your question. Detwiler says you are following the proper procedures for adding calcium chloride. Adding flake calcium chloride directly to the mixer can lead to discolored concrete as well as soft spots in the surface. Although the heat of reaction for calcium chloride dissolving in a bucket seems high, it wouldn't have much effect on a truckload of concrete. The calcium chloride solution doesn't have a shelf life, but be sure to agitate the solution before mixing into concrete.

Q. *What's the required mixing time and drum speed when an accelerator is added to ready-mixed concrete at the jobsite?*

A. ASTM C 94-96, "Standard Specification for Ready Mixed Concrete," doesn't address the jobsite addition of an accelerator. But you could probably follow the ASTM C 94 procedure for a single water addition that permits bringing the concrete to the specified slump. This procedure requires 30 drum revolutions at mixing speed (about 18 rpm or more), or as designated by the truck manufacturer. If you're using calcium chloride, it should be in a water solution, not in flake or pellet form.

To verify that mixing is adequate, observe slump uniformity as the concrete is discharged. If the first concrete down the chute is more fluid than concrete discharged later, the added calcium chloride solution might not be thoroughly mixed into the concrete.

Q. *I'm a poured wall concrete contractor and I use flowable concrete made with superplasticizers for many jobs. Sometimes there's a strong smell from the concrete. Also, I'm extremely allergic to formaldehyde and my skin breaks out when I work with flowable concrete. Do superplasticizers contain formaldehyde? Is that what I smell? Are there ways to make flowable concretes without using a superplasticizer?*

Adding chlorides for flatwork mix

Mixing time for jobsite-added accelerator

Alternatives for superplasticizers

A. Most superplasticizers do contain formaldehyde. They are either sulfonated melamine-formaldehyde condensates or sulfonated naphthalene-formaldehyde condensates. That may explain the allergic reaction you're having.

The smell is probably not formaldehyde. In concrete heavily dosed with naphthalene-based supers, however, you may be able to smell naphthalene. It smells like mothballs.

There are other options for producing flowable concrete. One way is to add both a conventional water reducer and a retarding admixture to the concrete. If you take this approach, watch out for setting problems, especially when slabs are poured. The concrete may set very slowly and delay finishing operations. On hot, dry days, the surface may crust over while the underlying concrete is still plastic. The top surface appears to be ready to finish but is spongy and will deform under the finisher's weight or the weight of a power trowel. This can cause a wavy slab surface.

Flowable concrete also can be produced by using a large dose of a conventional water reducer and adding enough accelerator to counteract the retardation effects. All water-reducers retard set to some extent and large doses can cause exceptionally long setting times unless an accelerator is added, too. Be aware that some accelerators contain calcium chloride and are prohibited on jobs where chlorides aren't permitted in the mix.

Causes for air content fluctuations

Q. *A ready mix truck arrives at the jobsite with 8 cubic yards of concrete. Travel time was 10 minutes. We discharge a yard, then take a slump and air test. The air content is 6% and slump is 2½ inches. We add 10 gallons of water to increase the slump, then discharge 5 more yards in about 15 minutes. If we test again, where will the slump and air be? And right after the water was added, what would the change in air content have been if we had run another test? Does temperature have an effect on air content? How about relative humidity?*

A. Adding water at the jobsite increases both slump and air content. The rule of thumb for slump is that 1 gallon of water added to a cubic yard of concrete increases slump by 1 inch. Air entraining agents form air bubbles by a frothing action. The wetter the concrete, the more frothing action you get. So up to a point, the higher the slump, the higher the air content. If the concrete gets too sloppy, however, some air may be lost.

Adding 10 gallons of water to 7 yards of concrete probably increased the air content slightly. Air content decreases after prolonged mixing or agitation but agitating for only 15 minutes after adding water isn't likely to reduce the air content or slump significantly.

Relative humidity doesn't affect air content, but temperature does. For a given dosage of air-entraining agent, as the concrete temperature rises, the air content decreases.

Permeability of air-entrained concrete

Q. *Why is air-entrained concrete less permeable than non-air-entrained concrete? Aren't the air voids permeable to water?*

A. Air-entraining agents reduce the amount of water needed to produce a given slump. At a fixed cement content, this lowers the water-cement ratio, which reduces permeability. Although entrained air bubbles increase the void content of hardened concrete, the air voids aren't interconnected. Thus, the higher void content doesn't appreciably increase permeability.

Reader response: The response to the question on permeability of air-entrained concrete neglected to mention a more important reason for the reduced permeability. The slight reduction in the water-cement ratio may have some bearing, but the entrained air bubbles make the mortar, or paste, in the mixture more cohesive, or sticky, according to finishers. Therefore, air-entrained concrete mixtures are more resistant to the settlement of aggregate particles. This results in reduced bleeding and fewer bleedwater channels through which water can enter the concrete.

The response correctly states that entrained air bubbles are not interconnected. But they are so small that water will not enter them under normal atmospheric pressure, so they don't affect permeability. Water can be forced into the small bubbles, however, under the increased pressure caused by the expansion of freezing water. Thus, the bubbles provide pressure relief.

Richard O. Albright, Arsee Engineers, Noblesville, Ind.

Q. *We accidentally added 6% calcium chloride by weight of cement to a truckload of concrete that went into a residential driveway placed during cold weather. Though the contractor had no trouble placing and finishing the concrete, he is concerned about low concrete strength. He heard that using more than 2% calcium chloride can cause strength loss at later ages. Is this true?*

A. Accelerators increase early strength at the expense of later strength, so a similar concrete without the accelerator would gradually reach a higher strength than the concrete with an accelerator.

If the contractor is concerned that strength at, say one year will be less than the 28-day strength, that situation is unlikely. We could find no data showing a 6% calcium-chloride dosage causing strength retrogression with age. But if there are no 28-day strength results because cylinders weren't made, you could core the driveway to determine strength of the in-place concrete. A 3000-psi or higher core strength should be adequate for a driveway in a cold climate. The driveway owner may not like the appearance of patched core holes, however.

Q. *During the heavy floods we experienced this past spring, our concrete sand was contaminated with lignite. We don't have enough room to build another sand stockpile, and I want to avoid hauling it back to the quarry and paying rehandling costs. Is there any way I can use the contaminated sand but avoid potential surface defects?*

A. Anti-washout admixtures (AWAs) might solve your problem. AWAs are water-soluble polymers that thicken the cement paste by physically binding the mixing water in concrete. They were developed for concrete placed under water. Most consist of microbial polysaccharides such as welan gums or polysaccharide derivatives such as hydroxpropyl methyl cellulose or hydroethyl cellulose.

AWAs decrease concrete slump, so you might need to add a high-range water reducer (HRWR) to counteract that effect without reducing strength or durability. Since the cost of both an AWA and HRWR may be too high, an alternative approach is to provide concrete with a maximum slump of 4 inches. Fewer lignite particles float to the surface in lower-slump concretes.

Q. *I'm confused by the terms manufacturers use to describe some of the admixtures in their "modified" cementitious repair mortars. I've heard the terms polymer-modified, latex-modified, acrylic-polymer, and epoxy-modified. What are the differences between these admixtures?*

A. All the mortars you mention fall under the general category of polymer-modified. The primary benefits of polymer modifiers are increased bond strength, reduced permeability, increased

Effects of a calcium-chloride overdose

AWAs salvage lignite-contaminated sand

Making sense of polymer modifiers

resistance to freezing and thawing, and increased flexural strength. The most common types of polymer modifiers are latexes. Although not a specific polymer, a latex is a polymer system consisting of very small spherical particles of high-molecular-weight polymers held in suspension in water by the use of surfactants. The most common polymers used in latexes for portland cement concrete are styrene butadiene, acrylics, and vinyl acetate copolymers. Styrene butadiene has been used extensively as an admixture in bridge and parking deck overlays. Acrylics and vinyl acetate copolymers often are used in manufacturers' prepackaged mortars.

Epoxy emulsions, although not classified as latexes, can be mixed with portland cement concrete to provide similar benefits. Epoxy emulsions are more expensive than most latex polymers but are gaining popularity as an admixture in cementitious bonding agents and rebar coatings.

Reference
ACI 548.3R-95, "Guide for the Use of Polymers in Concrete, American Concrete Institute," 1994.

Grout admixture for preplaced-aggregate concrete

Q. *We're involved in a project in which heavily reinforced concrete columns are being repaired with preplaced-aggregate concrete. One material specification calls for the grout to contain a "fluidifier." Are grout fluidifiers simply superplasticizers or are they some other type of admixture?*

A. ACI 212 "Chemical Admixtures for Concrete" does not consider grout fluidifiers to be superplasticizers, but lists them under the general heading of "Grouting admixtures." Grout fluidifiers usually contain water-reducing admixtures to increase flowability, a suspending agent to minimize bleeding, and an aluminum powder expansive agent to prevent bleedwater from collecting under the coarse aggregates. They also retard the set of the grout, increasing the time available for pumping the grout.

Using a shotcrete accelerator

Q. *My company is the structural engineering firm on a bridge repair project, and we have specified dry-mix shotcreting on several vertical and overhead surfaces. The project is running behind schedule and the contractor has recommended that we allow him to use a sodium carbonate accelerator in the shotcrete. He said that this will allow his nozzlemen to apply thicker, more frequent layers of shotcrete. We'd like to speed up the job, but we've heard that rapid-set accelerators reduce the compressive strength of shotcrete. Is this true?*

A. The effect of a sodium carbonate accelerator on the hardened properties of dry-mix shotcrete was reported in "Durability of Dry-Mix Shotcrete Containing Rapid-Set Accelerators," *ACI Materials Journal*, May-June 1992, pp. 259-262. As part of the study, a shotcrete mix containing 4% sodium carbonate accelerator by weight of cement was compared to a control shotcrete mix containing no accelerator. Both mixes had a cement-sand ratio of 1:4.

The 28-day compressive strength of the accelerated shotcrete was 54% lower than that of the control mix. Perhaps more importantly, the accelerated shotcrete had a reduced resistance to freezing and thawing (ASTM C 666). Concrete specimens are generally considered to have acceptable freeze-thaw resistance if they lose less than 5% of their weight after 300 freeze-thaw cycles. After 311 freeze-thaw cycles, the control mix had suffered a weight loss of only 0.9%. The accelerated shotcrete, however, lost 7.3% of its weight after 153 cycles and 17.1% after 210 cycles, at which time all three specimens had failed.

The study concluded that the reduction in freeze-thaw durability was caused by not only the reduced strength of the shotcrete, but also an increase in the porosity of the accelerated shotcrete. Apparently, the quick or flash setting of the shotcrete inhibits compaction.

Q. *My belief is that increasing the cement content in mixes designed for placement in conditions below 32° F is counterproductive to initial strength gain due to the additional hydration required to generate heat. We use a Type III cement and generally deliver concrete at 60° F at time of placement. Since hydration slows in colder temperatures, would a mix design with less cement and more water-reducing admixture produce higher one-day strength than a mix with higher cement content?*

A. Even if customers order concrete on days with below-freezing temperatures, ACI 306R, "Cold Weather Concreting" recommends that they maintain concrete at a minimum temperature for one or more days. For structures with minimum dimensions less than 12 inches, concrete must be placed and maintained at 55° F. The required duration of protection from early freezing depends on the type and amount of cement, whether an accelerating admixture is used and the service category of the structure.

With standard mixes placed and maintained at 55° F, contractors will certainly get enough cement hydration to generate heat. But to reduce the duration of the protection period, ACI 306R suggests that contractors request mixes that increase the rate at which the concrete gains strength by using a Type III cement, an accelerating admixture or an extra 100 pounds of cement per cubic yard. The extra cement reduces water-cement ratio, producing a higher strength at any age, and also supplies more heat of hydration at early ages.

Using less cement and more water-reducing admixtures, as you suggest, would be counterproductive because, even though you might get a lower water-cement ratio, the concrete wouldn't develop as much heat of hydration. A Type A water reducer might also cause undesirable set retardation in cold weather.

However, some Type E water-reducing, accelerating admixtures meeting the requirements of ASTM C 494 substantially improve the 24-hour strength of concrete maintained at 50° F. According to ACI 306R, the strength of concrete containing these admixtures approaches the strength obtained by using 2% calcium chloride and will be appreciably greater than for some, but not all, concretes made with Type III cements. The data also indicate that water-reducing accelerating admixtures may reduce setting times and produce substantial strength increases at all later ages. However, ACI 306R makes no mention of reducing cement content when these admixtures are used.

Q. *Can a frozen admixture still be used once thawed?*

A. It depends on the admixture. Most admixtures can be reconstituted after freezing. However, as a first step, call the admixture manufacturer and ask if homogeneity can be restored after freezing. If it can be restored, the next step is raising the admixture temperature to at least 50° F until it's completely thawed. Remix by mechanical agitation or with low-pressure air. Agitation should recirculate the mixture from the top to the bottom. Be careful when agitating by air because too much air causes admixtures to foam. Some admixture manufacturers warn against potential problems caused by agitating with air and recommend remixing their admixtures only by mechanical agitation.

Once you've thawed and reconstituted the admixture, the simplest way to test many admixtures is by trial and error. If a superplasticizer has frozen, for instance, batch the thawed material into a load of concrete and observe the effect on slump or water demand. You can test air-

Using water reducers in cold weather

Can admixtures be used after freezing?

entraining agents, accelerators, and other admixtures in a similar manner. This approach won't work for corrosion inhibitors and other admixtures that don't have a measurable effect on plastic concrete.

Alternatively, you can send a sample of the reconstituted admixture to the manufacturer. By testing for pH, specific gravity, residue by oven drying, or infrared analysis, the manufacturer can determine whether the thawed sample test results fall within manufacturing tolerances permitted by their quality control standards.

Air-entraining vs. gas-forming agents

Q. *Are gas-forming agents the same as air-entraining agents? If not, what are gas-forming agents and how are they used in concrete mixes?*

A. Gas-forming agents are not the same as air-entraining agents; each serves a different purpose. According to the Portland Cement Association, air entrainment is recommended for nearly all concretes, principally to improve their resistance to freezing. Air-entraining agents stabilize and incorporate within the concrete bubbles formed during the mixing process.

Gas-forming agents, on the other hand, are added for a different reason. Aluminum powder and other gas-forming materials are sometimes added to concrete and grout in very small quantities to cause a slight expansion prior to hardening. This may be of benefit where the complete grouting of a confined area is essential, such as under machine bases or in post-tensioning ducts of prestressed concrete.

These materials are also used in larger quantities to produce lightweight cellular concretes. The amount of expansion that occurs is dependent on the amount of gas-forming material used, the temperature of the fresh mixture, the alkali content of the cement, and other variables. Where the amount of expansion is critical, careful control of mixtures and temperatures must be exercised. Gas-forming agents will not overcome shrinkage after hardening caused by drying or carbonation.

Why is calcium chloride added to concrete?

Q. *An employee at our plant says calcium chloride is added to concrete as an antifreeze. However, I have always thought that it is added to help concrete gain strength faster so it can have a better chance to survive if it freezes the first night. Who is right?*

A. While it is true that calcium chloride lowers the freezing point of water, the amount of calcium chloride that can be used to accelerate concrete will not depress the freezing point enough to be noticeable. Therefore, that is not the reason for adding calcium chloride.

It is often said that the reason for adding calcium chloride in cold weather is to help concrete achieve a compressive strength of 500 psi before it freezes. This is a good rule of thumb. However, a 500-psi compressive strength represents a very small tensile strength; one that would be inadequate to resist the large internal forces generated when water freezes. Therefore, adequate strength is related to the reason for adding calcium chloride, but is not the real reason.

When portland cement hydrates, some of the mixing water gets used up by combining chemically with portland cement. By the time the concrete has achieved a compressive strength of about 500 psi, enough mixing water has combined with the portland cement so that the mixing water fills only 90% or less of the pours. That amount leaves enough space available for the water to expand in freezing without damaging the concrete. Therefore, the calcium chloride accelerates the reduction in water in the cement paste to a level where an early freeze will not completely disrupt the concrete. The same general principle applies to other accelerators used for cold-weather concreting.

Q. *What is the difference between liquid and solid calcium chloride accelerators for concrete?*

A. Calcium chloride is a crystalline solid that may be either anhydrous (water free) or dihydrate (combined with a small amount of water). It is available in flake and pellet form. So-called liquid calcium chloride is really a solution made by dissolving either anhydrous calcium chloride or any calcium chloride hydrate in water.

The solution usually contains somewhere between 28% and 42% anhydrous calcium chloride by weight of solution, depending on the specifications of the buyer. If flakes or pellets are to be used as an accelerator, they should be dissolved in water before being added to the concrete, so sometimes it is more convenient to buy a solution.

There is a danger if dry calcium chloride is added to concrete because it may not get thoroughly mixed into the batch. This could cause nonuniform setting or strength gain, and concrete cracking or localized corrosion of reinforcing steel if conditions promote corrosion.

Chapter 2

Aggregate Exposure– Sandblasting/Waterblasting

Q. *What is bushhammering and how is it used in concrete construction?*

A. Bushhammering is the use of impact tools to remove mortar and fracture aggregate at an exposed concrete surface. The amount of removal can vary significantly, depending on the application. Think of bushhammering as more aggressive than sandblasting and usually less severe than scabbling. Hand tools or bits for electric and pneumatic hammers are used, depending on the desired surface texture, the amount of surface to be tooled, and the location of the work. A special tool for bushhammering has several conical or pyramidal points, and resembles a meat tenderizer. Similar effects can be achieved by using pointed or chisel bits or needle scalers.

Bushhammering is done for various purposes. It can produce pleasing surface textures varying from a light scaling to a deep reveal of coarse aggregates. Other applications include removing fins, protrusions, or drips that can form on concrete during construction, or mineral deposits that can build up on concrete over time. Because it increases roughness, bushhammering also can be used to prepare concrete surfaces before patching.

Marble, calcite, and limestone aggregates are well-suited for concrete to be bushhammered for architectural purposes. Natural gravel can shatter or break out of the cement paste, while hard granite and quartz aggregates can break erratically or into the paste, rather than across the intended surface. Most structural concretes can be bushhammered. Use air-entrained low-slump mixes, properly placed and vibrated. Gap-graded and low-sand-content mixes are recommended for best appearance.

Concrete can be bushhammered after it reaches a strength of about 4000 psi and is at least two to three weeks old and surface dry. Be careful when working close to an edge or corner. It's a good idea to do a sample area for acceptance by the owner and specifier before beginning work. This can minimize misunderstandings about the intent or quality of the work.

Q. *Is there a standard mix to use for exposed aggregate flatwork if you do not want to seed the top of the concrete with a special aggregate?*

A. We are not aware of any standard mix of this type. You should design the mix or choose the mix with the desired end result in mind. This means that you will probably want to consider what maximum size of aggregate is best for the effect you want to achieve. Pea gravel (⅜-inch maximum) is a popular size for the purpose. It is often helpful to use a single size of coarse aggregate (for example, ¼ to ⅜ inch, or ⅜ to ¾ inch) rather than a graded aggregate. This tends to give a good appearance. If you use anything other than a standard mix, you should make sure it has the proper workability. If not, you may want to increase or decrease the sand content to make it more suitable for placing and finishing. All of this presupposes that the mix will be designed for adequate strength (preferably 4000 psi if for outdoor use in a freezing climate), will have a slump of 2 to 4 inches, and, if it is to be exposed to weather, will be air-entrained.

Wetblasting vs. waterblasting

Q. *A recent specification requires that we use "wetblasting" to remove an old coating from a concrete wall. Is wetblasting the same as waterblasting?*

A. No. Waterblasting uses water pumps as its power source and may or may not use abrasive for more aggressive cleaning. Wetblasting, on the other hand, uses compressed air and a standard blast machine to propel abrasive, with just enough water added at the nozzle to suppress dust. It produces the same hard-hitting force of dry abrasive blasting.

The simplest form of wetblasting uses a circular ring that fastens to the end of a blast nozzle. It directs streams of water into the air and abrasive as it leaves the nozzle. Typically, water is supplied from the nearest faucet or pumped from storage drums. These attachments spray water on the outside of the air and abrasive stream, leaving the center of the blast pattern dry. They work well where the local water supply has sufficient pressure for light dust suppression, but they have limitations.

Stripping old painted concrete can be extremely dusty, so you may want to consider using a wetblast injector rather than a simple wetblast attachment. An injector adaptor, inserted between the nozzle holder and the nozzle, introduces water through angled jets just before the abrasive flows through the nozzle. The wetblast injector uses a self-contained water pump to overcome the pressure in the blast hose. This thoroughly wets the abrasive particles before they leave the nozzle to provide more effective dust control.

Reference
Blast Off 2, Clemco Industries Corp., 1994.

Chapter 3

Colored Materials

Q. *We're bidding a job that requires a white concrete floor. Another bidder has proposed mixing white latex paint in with the concrete. Can this work?*

A. We checked with two experts in applications of latexes and other polymers in concrete, and neither has heard of this practice nor believes it would work. The latex used in paint is not compatible with portland cement, so adding latex paint to the concrete mix could make a gooey mess.

We recommend you stick to the proven methods of producing white concrete surfaces: use white portland cement, a white dry-shake, or apply a white surface coating.

Adding latex paint to concrete

Q. *We sometimes patch existing asphalt parking lots with concrete. On a couple projects, we saw-cut the area to be patched, removed the bad material, and replaced it with air-entrained concrete. This seems to be economical. The one drawback is the patches are so noticeable. Has anyone else done this before, and how can we color the concrete to make the patches less noticeable?*

A. Although *Concrete Construction* has published articles about whitetopping, we don't re-call anything specific on concrete used to patch asphalt lots.

As far as coloring, talk to your ready mix supplier about supplying concrete colored black. Water-dispersible carbon black can be used to make black concrete. Don't exceed 10 percent by weight of cement. There also are emulsified carbon black or black iron oxide pigments formulated for use in concrete. Add 2 to 5 pounds of pigment per sack of cement. To obtain uniform color, careful proportioning and extended mixing may be needed. Blending dry cement and color compounds can help reduce streaking.

Since you are concerned with durability in these paving applications, we recommend running an air content test on these mixes. Unmodified carbon black substantially reduces air content. Most carbon black intended for coloring concrete contains an admixture to offset the effect on air content.

Patching with black concrete

Q. *We wrote a job specification calling for application of a colored dry shake to 50,000 square feet of concrete floor. The contractor didn't read the specification and finished the concrete without applying the colored dry shake. Workers applied a membrane curing compound to the concrete. Is there any way to color the floor now so it doesn't have to be removed and replaced?*

Coloring flatwork after it's in place

A. Surface-applied stains can be used to color the concrete, but you won't get as deep a color as you would have achieved with the dry shake. To get the stain to penetrate, the contractor will have to remove the curing compound, either by using chemicals or abrasive blasting. Blasting will alter surface texture, but could open the surface so it accepts the stain more readily.

Regardless of the surface-preparation method used, the color of stained concrete won't be uniform. This nonuniformity is characteristic, and in some installations the softness or patina of the effect is valued.

We suggest that the contractor contact some concrete stain manufacturers, describe the problem and ask about staining options. He might try removing the curing compound in a small area, applying the stain and asking the owner if the resulting color is acceptable. (The curing compound manufacturer can provide information on how to remove the compound most efficiently.) Expect some variation in color over large areas because of batch-to-batch variations in concrete color.

Antiqued surface for existing concrete

Q. *How can we give an existing concrete surface an antiqued finish?*

A. That depends on what you mean by antiqued—here beauty is in the eye of the beholder. The question usually comes up with new concrete surfaces. Existing surfaces may be an even greater challenge, depending on the degree of exposure and extent of contaminants. Experiment with color and texture. Techniques such as scabbling, scarifying, shotblasting, and grinding can clean the surface and expose or fracture aggregates to reveal surface texture. Commercial concrete stains can achieve a range of colors. Check with a specific manufacturer about procedures to produce a mottled, rather than uniform, appearance. We've heard of washing with a solution of carbon black in water to achieve a look of "artificial weathering." You also might consider ferrous sulfate solutions to give a brown or buff color, or copper sulfate to give shades of green. Apply the solution to the concrete surface and let stand for about 15 minutes before washing off. Repeat until the desired effect is achieved.

Painting concrete in bright colors

Q. *Where can we get a deeply colored paint for concrete, one that is highly saturated with primary color? We need it for a multiuse building 600 feet tall. The owner wants several bright colors, not the stains, pastels, and earth tones commonly used on concrete. We're also concerned about resistance to cracking since there will be appreciable movement in a building of this height.*

A. Speaking in generic terms, we think you should use a two-component aliphatic urethane architectural coating. From a comparison of coatings made by James Kubanick several years ago (*Concrete Construction* May 1981, page 406) we learned that the aliphatic urethane coatings have excellent color and gloss stability under exterior exposure. They also have excellent solvent resistance, abrasion resistance, and hardness. Yet they are flexible, resist impact damage, and are not subject to long-term embrittlement.

The aliphatic urethanes also are available in deep colors. When we spoke with Dave Parish at General Polymers Corporation in Cincinnati, Ohio, one of the manufacturers of this type of coating, he assured us that they could match any color the architect or owner desired. Mr. Parish commented that some acrylic latex coatings would also give good service outdoors, but their color and gloss selections are limited.

Urethanes are sensitive to surface moisture and must be applied to a dry substrate. Carefully follow all the manufacturer's recommendations for safe handling and proper application. Do not deviate from the standard instructions without consulting a technical representative of the producer.

Q. *In general, what requirements pertain to the development of production-approval samples of architectural precast concrete?*

A. According to *Architectural Precast Concrete*, published by the Precast/Prestressed Concrete Institute, "Samples for architectural precast concrete should only be regarded as a standard for performance within the variations of workmanship and materials to be expected. If the color or appearance of the concrete is likely to vary significantly, samples showing the expected range of variation should be supplied. The concrete placement and consolidation method used for samples should be representative of the actual procedures used in the production of the element.

"When approved, the samples should form the basis of judgment for the purpose of accepting the appearance of architectural finishes. These samples also should establish the range of acceptability with respect to color and texture variations, surface defects, and overall appearance."

Q. *My salesman just accepted an exterior flatwork job that requires an integral color red concrete to complement the building's brickwork. The project requires about 300 yards, and the contractor will pour on three separate days. How should I plan to produce a consistently colored product?*

A. Good communication between you, the architect, and the contractor can eliminate many problems. Matching colors to a specific item can be difficult. Arrange for a test slab, not less than 3 cubic yards, to be poured well in advance of the construction target date. This up-front time is needed because the color of month-old concrete is different from that of newly placed concrete.

In your mix design preparation, be certain to use a minimum cement content of 470 pounds per cubic yard. Select color admixtures designed specifically for ready-mixed concrete. The most common method of coloring concrete is by adding premeasured disintegrating bags to the concrete mix. And be certain aggregate and cement from the same source are used throughout the job. Batching procedures must ensure identical mix proportions and water content from load to load.

Quality colored concrete is easy with good production practices.

Review the jobsite with the flatwork contractor. Extra care should be taken to prepare a well-drained and compacted subgrade. Review the type of curing compound the contractor plans to use. Review the selection with the color admixture manufacturer. A color-matched curing compound can help avoid uneven final coloring. Be certain the contractor doesn't cure with coverings such as burlap, plastic, or paper. These often cause discoloration. Also review order quantities so you can use full bags per truck. Don't try to batch partial bags. On the day of the pour, be certain truck drums are clean. Try to pour on days with similar ambient temperatures. Once the truck has completed mixing to a slump of 4 inches or less, minimize the amount of water added at the jobsite. Be certain all water quantities are exact so as not to dilute the color. Minimize the time between batching and placing, never exceeding 1 or 1½ hours.

A consistently colored mix is the result of thorough planning, good quality control, and proper placing techniques.

Chapter 4

Concrete Joints

Q. *Last summer, we built a slab-on-grade concrete parking lot. This winter, our finishing techniques are being blamed for damage caused by a plow during snow removal. The building supervisor is complaining about the joints that we hand-tooled. He's claiming the groover wasn't held perfectly flat, but was slanted, creating a slight ridge along the joint line. Naturally, when the plow comes through with the blade down, front wheels up, and all the pressure on the blade tip, the blade spalls off the joint edge. We don't think this is our problem. If it is, we'll be coming back after every winter to fix the spalls. What can we do to convince the building supervisor it's not our problem?*

A. First, take a look at the specification. We doubt if any specification addresses the allowable slant of the groover. Quite likely, you've met project specifications and the building owner is hoping you will provide life-long parking lot maintenance. We've never heard of any contractor being held responsible for damage due to snowplows.

Q. *We are having trouble with slabs cracking before we saw the control joints. Our procedure is to place the concrete slab, then come back the next morning to saw the joints. Before we start sawing, there are already some cracks in the slab. However, it's too early to start sawing the same day the concrete is placed, and we don't want to pay workers overtime to saw the joints in the middle of the night. The specifications say to saw the joints as soon as possible. Since the specifications don't tell us the exact time to saw the joints, we believe we've met the specifications. What do you think?*

A. We don't believe you have met the specifications, but there is a way to solve your problem. Remember, it's impossible for a specification to indicate an exact time to saw joints. Because of variations in weather and mix designs, the time to saw joints after concrete has been finished varies from around four hours in hot weather to as long as three days in cold weather. Engineers don't specify a time to saw joints because they don't know what time of year the concrete will be placed or even what time of day the contractor will decide to place the concrete.

Unfortunately, the time to saw joints is based on experience and trial and error. Experience with local weather conditions and mix designs provides a time window for sawcutting, but the exact time is still determined by trial and error. At the beginning of the time window for sawing, make a sawcut in the slab. If the edges of the sawcut ravel, dislodging aggregate, then it's too early to saw. Continue to saw at regular time intervals until the cut edges show only minimal raveling. When this happens, the slab is ready for sawcutting.

Sometimes concrete cracks before the saw can cut the joint without causing raveling. In these cases, consider using a lightweight saw that allows you to cut within a few hours after final concrete finishing. Using this equipment, you can place concrete in the morning, then cut the slab during regular working hours the same day. This eliminates costly overtime and minimizes the waiting time associated with the trial-and-error procedure used to determine sawcut timing.

Grooved joints blamed for plow damage

When to saw control joints

Gasoline-powered saws with 8- or 10-inch-diameter blades are used for runways, pavements, and industrial floors. They feature automatic depth control and sawing speeds from 7 to 30 feet per minute. Sawcut depth is up to 1⅜ inch and sawcut widths range from a ⅛-inch straight groove to a ½-inch T-cut. Electric-powered lightweight models use a smaller-diameter saw blade and cut notches ¾ to 1 inch deep at sawing speeds of 4 to 11 feet per minute. For more information on these saws, see "Roundup of Early-Cut Concrete Saws," *Concrete Construction*, May 1998, pages 439-441.

Joint detail where asphalt pavement meets a concrete slab

Q. *I'm designing a new fleet shop for the company where I work. Tractor trailers will drive out of the fleet shop, which has a concrete floor, onto an asphalt parking area. My concern is load transfer from the concrete to the asphalt. Should the asphalt be thickened at the joint? Would dowel bars be beneficial? I need a long-lasting, durable joint.*

A. You won't be able to transfer load directly from a concrete slab to a flexible pavement. If you're stuck with an asphalt parking lot, the best approach is to pave a concrete apron around the building, with a slope that will ensure adequate drainage away from the building. Thoroughly compact the subgrade and base material adjacent to the apron, and slope the asphalt lot to minimize water penetration at the joint. Pay particular attention to asphalt compaction at the joint.

Expansion joints needed for white-topped parking lot?

Q. *I've recently been awarded a contract to whitetop an existing asphalt parking lot in Pennsylvania. One section of the lot is 80x200 feet. Should there be an expansion joint in a section of lot this large?*

A. You'll need isolation joints (also called expansion joints) where the lot meets any building, light pedestal, or other fixed object. According to Dale Diulus of Phoenix Cement Company, however, you won't need any expansion joints in the 80x200-foot section. Contraction joints at 8- to 10-foot intervals should accommodate any slab movement caused by temperature changes and drying shrinkage. Sealing the contraction joints will help keep out dirt and debris that might restrict movement at the joint.

Timing joint sawing in cold weather

Q. *We placed a 46x90-foot concrete floor slab that's 5 inches thick. The plans called for sawed control joints at 15- to 18-foot spacings. The slab was placed during cold weather so we used heated mixing water and 2% calcium chloride in the concrete. Concrete temperature at delivery was 60° F and the air temperature was 35° F.*

We finished the pour yesterday at 5 p.m. and covered the slab with plastic sheeting and 2-inch-thick insulated pads. The temperature stayed at or slightly below freezing last night and today it's still in that range with sleet and snow forecast. That means we won't be able to saw joints until tomorrow or even later and I'm concerned about random cracking. What do you suggest doing?

A. Under the conditions you had to work with, random cracking is likely to occur despite your best efforts. The cold weather is working against you in two ways. Concrete gains strength more slowly in cold weather. Thus you have to wait longer before cutting so that joint edges don't ravel. And as the slab cools it wants to contract. This restrained movement causes tensile stresses in the concrete. Because of these stresses, the concrete may crack before you can saw it.

On the other hand, the cold weather may work for you. Because concrete gains strength slowly in cold weather, it's not as stiff and may creep more under the tensile stress. We suggest getting on the slab as soon as possible to see if you can saw without excessive raveling. If it's possible, saw the joint that's nearest the middle of the 90-foot length. Sawing contractor Rick Younger advises that you use a diamond blade and cut the joint at least 2 inches deep. This will divide the slab into two nearly square sections and may also release some of the stress that could cause random cracking. Cut the remaining joints as soon as weather permits.

Q. *When a concrete wall is built on a concrete base slab and both are designed with construction joints that include a waterstop, shouldn't the joints match up? We built a water tank with 25-foot-high walls on a 3-foot-thick base mat that's about 50x50 feet. Plans called for construction joints with a waterstop at 25-foot spacings in the base mat. Plans also showed contraction joints with waterstops in the wall, but the wall joints were offset as much as 4 feet from the base-slab joints.*

Hairline cracks developed in the wall as shown in the sketch, and water came through the cracks when the tank was leak-tested. We had to epoxy inject the cracks to stop the leakage. I believe this wouldn't have happened if the joints had matched up. Is there an American Concrete Institute (ACI) or Portland Cement Association (PCA) document that says joints in a base slab should match with joints in a wall?

A. We checked ACI 350R-89, "Environmental Engineering Concrete Structures," ACI 301-96, "Specifications for Structural Concrete," and PCA's *Building Movements and Joints* but found no mention of a need to match joint locations. However, a 1981 article on joints in sanitary-engineering structures deals with this subject (Ref. 1). Author Roger Wood says that any joint in a structure should preferably go through the entire structure in one plane. He marks the joints on the first conceptual plans with big, heavy black lines, creating a visual effect that resembles a model of the structure after it's hit with a meat cleaver. Each slice made in the model is a joint location.

Wood explains his rationale for keeping the joints in one plane as follows: "Each joint is, to some degree, a discontinuity in the structure. Movement may take place at a joint. I can protect the joint from leakage due to movement by waterstops and sealants. If the joints are not in line and movement occurs in a portion of the joint it will probably tear a crack in the concrete above until the crack intercepts another joint. There is no waterstop or sealant present at cracks to prevent leakage."

This discussion seems to describe what happened to the tank you built. And Wood clearly agrees with you that joints in the base slab and walls should match up.

Reference

1. R. H. Wood, "Joints in Sanitary Engineering Structures," *Concrete International*, April 1981, p. 54.

Reader response: Concerning the question on whether or not to line up construction joints that include a waterstop (p. 630), you said that after checking several American Concrete Institute (ACI) and Portland Cement Association (PCA) documents, you found no mention of a need to match joint locations. However, ACI 224.3R-95, "Joints in Concrete Construction" (chapter 3, section 3.3.2, second paragraph), states: "Contraction and expansion joints within a structure should pass through the entire structure in one plane (Wood 1981). If the joints are not aligned, movement at a joint may induce cracking in an unjointed portion of the structure until the crack intercepts another joint."

Although this paragraph is taken from the chapter on joints in buildings, the topic of the referenced paper by Roger Wood is sanitary-engineering structures.

Grant T. Halvorsen, Structural Engineering and Concrete Materials, Wheaton, Ill.

Expansion joints not needed in sidewalk

Q. *I have been contracted to build 2,000 linear feet of 5-foot-wide, 4-inch-thick sidewalk on a 4-inch gravel base. The specifications call for expansion joints every 20 feet. I contend that expansion joints are needed at a minimum of every 50 feet and where the walk will abut existing structures. Control joints will be cut every 5 feet. Am I right in suggesting expansion joints every 50 feet?*

A. According to several industry sources, expansion joints, even at 50-foot intervals, are not necessary for the sidewalk project. Long stretches of concrete do not require intermediate expansion joints. Proper use of contraction joints at short spacings of 5 feet will allow for proper movement of the individual sidewalk slabs. Due to normal shrinkage of the concrete after placement, the slab probably will never expand and become larger than it is at the time of placement.

Isolation joints, however, will be needed where the walk will abut existing structures. According to ACI 332, "Guide to Residential Cast-in-Place Concrete Construction," isolation joints, sometimes called expansion joints, are only necessary to separate the sidewalk from a fixed or different concrete structure. Examples include separating the sidewalk from lamp posts, hydrants, footings, buildings, driveways, and curbs.

Tooled contraction joints every 6 to 8 feet weren't enough to prevent this golf cart path from cracking midpanel. Now the contractor installs fiber expansion board in the paths every 250 feet to prevent this problem.

Editor's note: In response to a Problem Clinic question in the January 1995 issue of *Concrete Construction* regarding the need for expansion joints in sidewalks, we said that sidewalks do not require expansion joints if properly spaced contraction joints are used. However, Bob Banka of Mid-South Concrete Path Paving, Kingston Springs, Tenn., has run into problems with golf cart paths buckling when only contraction joints are used. Here's his solution.

From our experience installing golf cart paths, which are basically long sidewalks, we have found that we need to install fiber expansion board material about every 250 feet. Before we started this practice, we had a few occasions where our cart paths would buckle during hot summer days. Sometimes the concrete would rise 6 inches.

We tool our joints 1 inch deep every 6 or 8 feet, depending on the width of the cart path. Sometimes a panel would rise and crack at the joint, and sometimes, as the photo shows, it would crack midpanel. The problem seems to affect our cart paths placed on sandy soil or in areas that have a high water table.

To be safe, we now install 1 inch of fiber expansion board material with three dowels (one in the center and the other two about 6 inches from each edge of the path). Since we started this practice, we have eliminated the buckling.

We are not sure why the concrete would sometimes buckle. The only reason we have been able to come up with is that the fiber reinforcement we use holds the joints together so tightly it does not allow the slabs to move independently.

Q. *We used an early-cut saw to joint a concrete floor immediately after finishing. The floor is 6 inches thick and the sawcut depth is ¾ inch. Although the floor has no random cracks, the engineer says we now have to chase the joints with a conventional saw so the cuts are 1½ inches deep (one-quarter the slab thickness). Is this necessary? We think it's a waste of time and money.*

Required joint depth for early-cut saws

A. A Portland Cement Association publication, *Concrete Technology Today* (November 1995), cites research showing that sawcut timing is much more important than sawcut depth in controlling random cracking of pavements. The researchers, in a field investigation by the Texas Highway Department, monitored joint formation and crack control for 13-inch-thick plain concrete pavement test sections placed directly on subgrade soils. Variables included differing concrete mixtures, coarse-aggregate types, curing methods and sawcutting techniques.

Joints in the test sections were sawed at 15-foot intervals using two different methods—conventional water-cooled sawcutting to a 3-inch depth and early-age sawcutting to a 1-inch depth (without cooling water). The early-age sawcutting was typically done less than three hours after concrete placement.

Observations over a 10-month period showed that of all the transverse cracks that developed, only two occurred between sawcut joints, and both of these uncontrolled cracks originated at re-entrant corners for inlet drainage structures. This investigation seems to indicate that the shallower depth of early-cut joints doesn't reduce their effectiveness in controlling random cracking.

Original research results are reported in Transportation Research Record 1449, published by the Transportation Research Board, Washington, D.C. Call 202-334-2934 for ordering information. You can buy a copy of the *Concrete Technology Today* article by calling 800-868-6733 and asking for PL953.01.

Editor's note: This item originally appeared in *Troubleshooting Newsletter,* published by the American Society for Concrete Construction. For more information about this quarterly newsletter, call ASCC at 800-877-2753.

Reader response: In the June 1997 Problem Clinic (p. 529) the reader wanted to know if it was necessary to chase ¾-inch-deep sawcuts, made by an early-cut saw, to a depth of 1½ inches (one-quarter the slab thickness).

I believe that many slabs on grade have large variations in slab thickness due to poor sub-grade preparation and construction procedures, resulting in planes of weakness requiring joints that exceed the depth of early-age sawcuts.

You quoted results of a research project by the Texas Highway Department, where early-age sawcuts gave excellent results. I would guess that the project had good slab-thickness control, which is not found on many projects. I recommend that all slabs on grade be cut to one-quarter the slab thickness.

Larry J. Asel, Conco Companies, Springfield, Mo.

The information on sawcut depth in the *Concrete Construction* June 1997 Problem Clinic is wrong. About every 20 years, the United States has a rash of off-center longitudinal cracking in concrete pavements after a cool fall. In subsequent field surveys, it was found that longitudinal joints were sawed to a depth of less than one-quarter the pavement thickness. Most states now specify that the depth of the longitudinal joint be one-quarter the slab thickness plus ¼ inch, and many recommend a depth of one-third the thickness.

I assume that in the research you referenced, the concrete was *not* tested by nature. Contractors who saw joints to a depth of less than one-quarter the slab thickness will experience problems if the weather turns unusually cool after concrete placement. The thermal shock will cause cracking, although the cracks may not be visible until months later.

Thomas J. Pasko Jr., Director, Office of Advanced Research, Turner-Fairbank Highway Research Center, McLean, Va.

Isolation joint depth

Q. *Do isolation joints need to be cut completely through a slab? For instance, will a 2-inch-deep cut allow vertical movement in a 5-inch-thick slab?*

A. Isolation joints are used to permit both horizontal and vertical movement at adjoining parts of a structure. To permit vertical movement, they should pass all the way through the concrete cross section. Otherwise, aggregate interlock would restrict vertical movement.

Rather than being sawed, isolation joints are often formed with isolation-joint material such as premolded fiber strips that extend the full depth of the slab. Columns are often isolated from floor slabs with circular or diamond-shaped box outs.

Reader response: Your reply is for the most part correct. But it contributes to the general confusion caused by the fuzzy nomenclature related to joints. Isolation joints must be formed or possibly cast around a full-thickness blockout. They can't be cut or tooled to achieve a reliable and complete separation between concrete elements. The response should have been more firm on this point.

Grant T. Halvorsen, Structural Engineering and Concrete Materials, Wheaton, Ill.

Movement in parking-deck expansion joints

Q. *Do expansion joints between parking-deck slabs allow for differential vertical slab movement when one slab carries much greater loads—due to more vehicles—than the other slab?*

A. Joint detailing determines whether or not a parking-deck expansion joint deflects. In cast-in-place construction, the joint often occurs over a beam, permitting horizontal but not vertical movement. Differential vertical movement is undesirable for some joint-sealing systems—such as those with cover plates—because the deflection would interfere with the seal.

Also, if the deck will be snowplowed, the plow blade would hit the raised portion of the deck if the other side were allowed to deflect.

In precast/prestressed parking decks built with double tees, there may be a shear key to prevent deflection of one double tee relative to the adjacent one. Or beams may be close enough to the joint to prohibit excessive movement.

Some joint-sealing systems permit both vertical and horizontal movements, and if there is no other reason (such as snowplowing) for not permitting vertical movement, some vertical movement may be acceptable.

Pinwheel joints

Q. *I've heard the term* pinwheel joint *used in connection with concrete floors. What is a pinwheel joint?*

A. The drawing below shows a plan view of this type of joint, named for its resemblance to a pinwheel. Since this joint type eliminates one step in construction, pinwheel joints sometimes replace the diamond or circular column-isolation joints typically used in industrial floors. It reportedly also produces a cleaner detail at the column.

The fibrous isolation-joint material is clamped to the column before floor construction. As the concrete floor is poured, workers also place concrete within the column H-section to support the isolation-joint material.

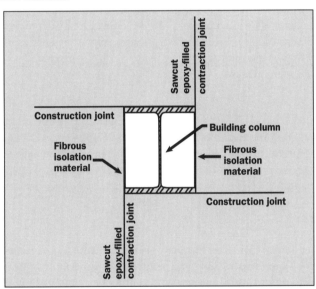

Repairing deteriorated expansion joints

Q. *We've been asked to repair joints in the concrete sidewalks at a shopping center. Every other joint was formed with expansion joint material that has deteriorated. It has shrunk to about an inch below the 1-inch-wide joint and is a tripping hazard for women with high-heel shoes. The material remaining in the joint is waterlogged.*

What's the best procedure for repairing the joints? I'm worried that joints filled with a soft sealant will still be a tripping hazard for high heels.

A. We'd recommend doing some test sections first and trying several pourable sealants. Contact sealant manufacturers and ask them to recommend products for the application. Remove all the waterlogged material from several of the joints. Put in a backer rod so sealant depth is half the joint width. Then fill different joints with the sealants to be tested. Observe the results when a test subject walks on the sealed joint with high-heel shoes during warm weather. Pick the best product based on your observations.

Injecting leaking cold joints

Q. *An engineer at a local water department recently asked us if we could epoxy-inject leaking cold joints in one of their water tanks. We've never injected cold joints, and we couldn't determine the exact places along the cold joint that were leaking. Can leaking cold joints be injected? Where do we set the ports?*

A. According to John Trout and Gary Hayes of Lily Corp., cold joints, unlike cracks, are not *voids* in the concrete—they are merely *seams*, containing no appreciable voids. Therefore, as a general rule, they cannot be injected. However, many cold joints do have small voids in areas where the concrete is not completely consolidated. Also, if water is passing through the cold joint, you can safely conclude that the concrete is at least porous enough to be saturated with low-viscosity epoxy resin injected under pressure.

To find the sources of the leaks, use compressed air to blow the concrete surface dry around the cold joint. Then watch to see what areas wet out first—you'll want to place your injection ports there. Drill ½-inch-diameter, 1-inch-deep holes using a wet diamond bit. Don't use a dry bit because it will impact the wet dust into the void and block the penetration of epoxy. Also, you don't have to apply a seal to the whole cold joint; sealing an inch or so on each side of the port should suffice. After you've placed the ports, drain the tanks below the level of the cold joint and wait a few days for the concrete to lose some of its moisture. Then, using a low-viscosity epoxy (200 cps or less), inject the cracks at as high a pressure as possible for as long as the pot life of the epoxy permits. If epoxy emerges from the inside face of the concrete, you'll see a noticeable drop in pressure.

Sealing industrial floor joints

Q. *Our company performs industrial floor repairs, and I have a long-standing argument with a co-worker about the proper way to seal joints. I've read that joint sealants should be twice as wide as they are deep so that they don't tear when the joint opens and closes. My friend disagrees. He often pours the sealant into narrow sawcuts, making the depth of the sealant far greater than the width. Which method is right?*

A. The proper joint sealing technique depends on the service conditions of the floor. The joint sealant shape you recommend is often correct for exterior concrete that is exposed to changing temperatures because the joints open and close when the concrete expands and contracts. For these applications, a flexible sealant is commonly used. The primary function of the sealant is to prevent water and incompressible materials from entering the joint.

Many industrial floor slabs, however, do not undergo much thermal movement because they are maintained at a fairly constant temperature. In addition, older floors have undergone most of their drying shrinkage, so movement at the joint is minimal. In these cases, a wide, shallow joint sealant shape is not required.

Many industrial floor joints are subjected to abuse from vehicles equipped with small, hard wheels that chip away at the joint faces. The primary purpose of the joint sealant (more properly called a joint filler) is to support the slab edges to prevent spalling. The recommended joint filler for these applications is a semirigid epoxy.

Q. *We've been asked to repair deteriorated joints in a 2-year-old concrete floor. The floor is 8 inches thick and is placed on a 4-inch-thick crushed stone base. The 3000-psi concrete was treated with 2 pounds per square foot of a dry shake hardener. Keyed joints are spaced 50 feet apart and there are two layers of wire mesh in the floor.*

Here's the problem. The keyed joints have opened up as much as ⅜ to ⁷⁄₁₆ inch. Then under hard-wheel dolly traffic, the top edge of the female side of the keyed joint has broken off. There's also spalling on both sides of the joint.

To repair it, I'm proposing that we saw out a 3-foot strip of concrete centered on the joint. Then at each sawed face we'll drill holes to accept smooth dowel bars and epoxy one end of the dowel while leaving the other end free to move within the concrete. We'll place a low-slump repair concrete and saw a joint above both rows of dowels. The joint will be filled with a semirigid epoxy. Does this sound like a good approach?

A. We asked Steve Metzger of Metzger-McGuire for a critique of the proposed repair. He thought the basic approach was sound but suggested cutting only one joint at the center of the repair section. Damage from hard-wheeled traffic is most likely at the joint. The more joints, the more damage is likely to occur. Since most of the floor shrinkage has probably occurred, the sawed joint isn't likely to open very wide and aggregate interlock should provide the needed load transfer. Metzger suggests a narrow-width sawed joint that can be achieved by using saws that make a cut immediately after finishing. The narrower the joint, the less likely that you'll get further joint distress.

Q. *On a concrete overlay job, the contractor did a poor job of installing a joint sealant and the sealant failed soon after it was applied. We were hired to do the repair but cold weather has set in. We're thinking of temporarily sealing the joint with a backer rod and hot-pour asphalt, then coming back in the spring to do the job right. Does this sound like a workable plan?*

A. If cold weather has already caused the joint to open, you might be able to keep out dirt and other debris with only a backer rod. Then you won't have to clean out asphalt before applying the new joint sealant. Use a closed-cell polyethylene foam backer rod. It won't absorb water and is less compressible than a polyurethane foam.

Q. *As part of the seismic upgrade of a 20-million-gallon drinking water reservoir, we are replacing all sealants in the floor and wall control joints. A typical sealant reservoir is 1½ inches deep, ½ inch wide at the bottom, and ¾ inch wide at the concrete surface. The tank has been drained, but the concrete is damp and can't be dried economically. What joint sealants can be used to replace the existing sealants?*

A. The results of tests by the U.S. Bureau of Reclamation may help. Released in December 1993, Report No. R-93-18, "Elastomeric Canal Sealants: Application to Wet Concrete" describes the performance of 16 sealants applied in both damp and underwater conditions. Test specimens were prepared in accordance with ASTM C 719, "Standard Test Method for Adhesion and Cohesion of Elastomeric Joint Sealants Under Cyclic Movement (Hockman Cycle)." Twelve specimens were prepared for each sealant: four on dry concrete, four on damp concrete, and four on concrete placed underwater. After curing, half the sealant specimens were tested for ultimate tensile strength and elongation. The other half were tested for tensile mod-

Floor joint repair

Can joint sealing be put on hold?

Joint sealants for damp surfaces

ulus of elasticity at 25% and 50% elongation. These modulus specimens were then placed underwater to simulate service conditions and retested for tensile modulus every two to four weeks for 26 weeks. Twelve of the 16 sealants received high or moderate scores when applied to damp concrete. Four of the sealants received high or moderate scores when applied underwater. For copies of the test report, contact Jay Swihart at the Bureau of Reclamation (303-445-2397; fax: 303.236.4679); e-mail: jswihart@usbr.gov

Repairing industrial floor joints

Q. *The joints in our 4-year-old warehouse floor have spalled to different degrees—from minor edge chipping to more major ruts that are affecting forklift traffic. Can we just prepare the deteriorated areas, then fill the gap with a sealant or other material? Or do we need to reconstruct the slab edges and create a new joint?*

A. Before repairing the joint, you should first determine if the deterioration is caused by a loss of subgrade support at the slab edges. If it is, simply repairing the spalled joint will probably provide only a temporary solution.

Loss of subgrade support (often caused by slab curling) causes the slab edge to deflect when a forklift runs over the joint. The forklift wheels then impact the edge of the adjacent slab and spall the concrete (Figure 1).

You often can detect slab deflections by simply standing on the joint while a forklift runs over it. To measure the exact deflection, consider using laser levels or dial gauges supported by Benkelman beams. Various rules of thumb can determine how much slab deflection is too much. References 1, 2 and 3 describe a few of these guidelines.

Consider undersealing the joint to stabilize excessive slab deflections. Reference 4 contains excellent information on undersealing.

Once you address the slab deflections, repair the joint spalls. Steve Metzger of Metzger/McGuire Co. (Reference 5) recommends repairing joint spalls less than 1³⁄₁₆ inches wide with an aggregate-filled semirigid epoxy joint filler (Figure 2). Use a semirigid epoxy with a minimum Shore hardness of A80 or D50 (ASTM D 2240). Add 2½ parts silica sand to one part epoxy to give the sealant more deflection resistance. If the repair area is less than ½ inch wide, don't add aggregate to the epoxy.

If spalled areas are wider than 1¼ inches, Metzger recommends creating a new joint by reconstructing the slab edges with a rigid epoxy mortar. After sawcutting and removing loose debris, insert a ⅛-inch-thick plastic strip into the joint. This will serve as a removable form. Mix a batch of neat epoxy and apply a prime coat to the bottom and sides of the repair area. Then add the recommended amount of sand to the liquid epoxy, mix thoroughly and apply the material to the repair area. Trowel the surface flush or leave it slightly high.

After the mortar has cured, remove the form, clean the joint and install a neat semirigid epoxy joint filler. After the filler has cured, sand or grind the area flush to the surrounding floor.

References

1. Ron Bartelstein and Ed Weiner, "Repairing Industrial Floors," *Concrete Repair Digest*, Spring 1990, pp. 7-12.

2. Peter J. Nussbaum, "Repairing Joints and Cracks in Industrial Floors," *Concrete Repair Digest*, April/May 1992, pp. 49-52.

3. Bruce A. Suprenant, "Repairing Curled Slabs," *Concrete Repair Digest*, December 1996/January 1997, pp. 296-301.

4. John G. Meyers, "Stabilizing Slab Deflection in Industrial Floors," *Concrete Repair Digest*, April/May 1992, pp. 57-60.

5. Steve Metzger, "A Closer Look at Industrial Floor Joints," *Concrete Repair Digest*, February/March 1996, pp. 9-14.

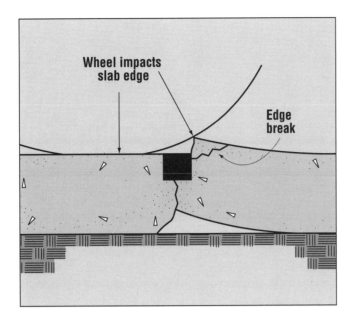

Figure 1. Loss of subgrade support (often caused by slab curling) causes the slab edge to deflect when a forklift runs over the joint. The forklift wheels then impact the edge of the adjacent slab and spall the concrete.

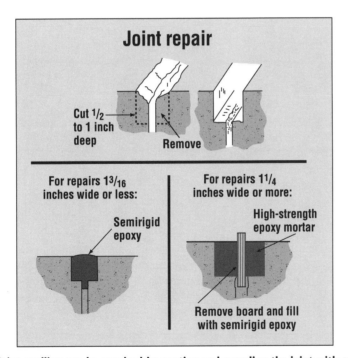

Figure 2. Minor joint spalling can be repaired by routing and resealing the joint with an aggregate-filled semirigid epoxy. To repair more extensive spalling, reconstruct the joint edges with a rigid epoxy mortar and reseal the joint with a neat semirigid epoxy.

Chapter 5

Crack Repair

Q. *About two months ago, we injected a crack in a residential foundation. The crack was about ⅛ inch wide at the surface. We weren't sure if we could control resin drainage, so we used an epoxy gel rather than a lower modulus material. Since the wall was 8 inches thick, we installed the entry ports 8 inches apart, capped the crack, and injected the resin. We started injecting at the bottom of the crack and worked our way up, plugging the port when epoxy flowed out the next port. Everything proceeded normally. The owner called recently saying that another crack had developed next to the old crack. Conventional wisdom is that the crack we injected was a moving crack and shouldn't have been injected. But I'm sure the crack was not moving. Is there any other explanation?*

A. We spoke to John Trout of Lily Corp. and he says he sees this problem often. The crack, ⅛ inch at the surface, probably narrowed at greater depths. Most likely, the epoxy gel you injected did not flow through the whole 8-inch-thick crack before appearing at the next port. If a crack is not completely filled, a plane of weakness is formed (a control joint, in effect) at that spot. Any stress on the wall is likely to induce a crack there. The new crack you see isn't really new at all. It's where the uninjected portion of the old crack resurfaced (see figure).

To increase your chance of completely filling a crack, Trout recommends that you start injecting a crack at its widest point regardless of where that point is on the wall. Stay on the wider part until you get back pressure, not just until you get bleed at the adjacent port. If the port starts bleeding before back pressure builds up, plug it and continue dispensing until the machine stalls out.

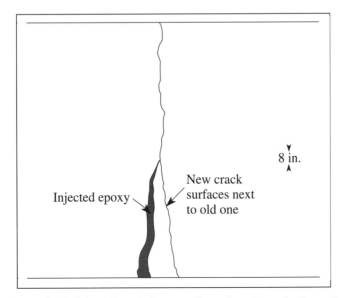

8 in.

Injected epoxy

New crack surfaces next to old one

Incompletely injected crack forms a plane of weakness in the wall.
Any stresses on the wall cause a crack to form next to the original crack.

Epoxy specifications for crack injection

Q. *We're bidding on a crack injection job where the requirements for the injection epoxy resin are specified. The specs require a compressive strength of 15,000 psi, tensile strength of 8000 psi, and slant shear strength of 5000 psi. We're having trouble finding an epoxy resin that meets these specifications. Any suggestions? The concrete we're repairing has a 3500-psi compressive strength.*

A. The required strengths are probably unnecessarily high. Ray Schutz, a consultant on adhesives, suggests that you try to get the specification changed so epoxy requirements are in line with those in ASTM C 881, "Standard Specification for Epoxy-Resin-Base Bonding Systems for Concrete." Schutz says it's likely that an epoxy resin with 15,000-psi compressive strength will be too brittle.

Removing capping material

Q. *We're working on a job requiring epoxy injection of architectural concrete. The specifications require visual approval from the architect. We're concerned that removing the capping material will mar the appearance of the concrete. What type of capping material is easiest to remove? How should we remove it?*

A. Peter Emmons, president of Structural Preservation Systems, says that you can use an epoxy capping material, but that you should burn rather than grind it off. Remove any remaining discoloration with a wet sandblaster.

Bob Gaul, president of Construction Polymer Technologies, suggests two alternatives to an epoxy capping material: hot wax or cementitious capping material.

Hot wax can be removed from flat surfaces with a putty knife. Light sandblasting or rubbing with an abrasive cloth may be necessary to remove any glossy look left behind. Hot wax must be applied to dry concrete and should not be used for injection pressures more than about 80 psi.

Prepackaged, fast-setting, cementitious mortars that are commonly used for thin patches also may work. Trowel the mortar into and around the crack, wait for it to set, then inject the epoxy. Afterwards, the mortar outside the crack can be removed with sandpaper. Since some mortar is left in the crack, consider coloring the mortar to match the concrete. When using cementitious capping material, injection pressure also is limited to about 80 psi.

Identifying moving cracks

Q. *We've been asked to repair a foundation wall that has a vertical crack in it. I know that if the crack is not moving, it can be injected with epoxy. I don't think it's a moving crack, but I'm not sure. What's an easy way to find out?*

A. The easiest way to tell if a crack is moving is to measure the crack width with a crack comparator or graduated magnifying device (Figure 1) at regular time intervals, every day or every week. A magnifying device is more accurate, but costs $50 to $100. The crack comparator is easy to use, sufficiently accurate for most jobs, and usually is supplied free by firms specializing in failure investigations. Record both crack width and date of measurement. Draw a line across the crack to mark where you measured it and always measure the crack width at the same location. Do this at three or four places along the crack.

Another inexpensive method is to patch the crack with plaster (Figure 2). Use hot water to speed the set of the patch. Note the date the patch was placed and inspect it at regular intervals to see if it has cracked. A cracked patch shows that the crack is active. Use a nonshrink patching material so the patch doesn't crack from drying shrinkage.

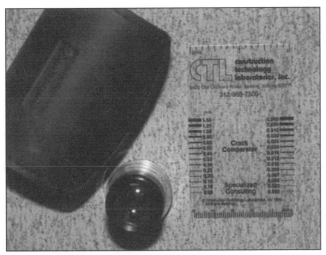

Figure 1. A graduated magnifying device (left) and a crack comparator (right) easily measure crack width.

Figure 2. When placed over a moving crack, a spot patch of plaster will crack.

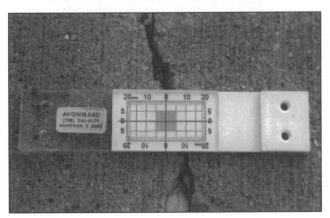

Figure 3. This crack monitor detects the amount and direction of crack movement.

A two-piece crack monitor can also be used to detect crack movement (Figure 3). One of the two plastic pieces has red cross hairs; the other piece has a grid system with a zero mark at the center. The two pieces come taped together so that the intersection of the red cross hairs on the one piece coincides with the zero mark on the grid of the other piece.

The monitor is laid across a crack and each end is attached to the wall with epoxy or a fast-setting glue. After the adhesive cures (about 15 minutes for fast-setting glue; about 24 hours for epoxy), the tape holding the two pieces together is cut. If either side of the crack moves, one of the plastic pieces moves. The red cross hairs, originally at zero, slide over the grid system indicating the amount of vertical and horizontal crack movement. Both the amount and direction of movement can be observed. Usually the monitor is checked at the same time every day.

Crack injection: epoxy or polyurethane?

Q. *We're working on a large remodeling job requiring the repair of several cracks and a leaky control joint in the poured concrete basement walls. The architect is requesting we inject both the cracks and control joint. The control joint is actively leaking. The cracks are not leaking, but some are wet. How do we decide whether to inject a polyurethane or an epoxy? Can we inject a crack or control joint that is wet or leaking water?*

A. To help decide whether to use a polyurethane or an epoxy injection material, ask the following questions:

• *Is the crack dormant, active, or moving?*

A dormant crack remains constant in width, whereas an active, or live, crack opens and closes as the structure is loaded or thermal changes take place. A growing crack increases in width because the original reason for its formation persists. See *Concrete Repair Digest*, October/November 1992, p. 223, for methods of identifying moving cracks.

Epoxies are relatively rigid and generally not suitable for injecting active or moving cracks. Urethanes are flexible and will tolerate typical crack movements. George Matson, Allied Concrete Repair, Plainfield, Ill., says he uses a polyurethane almost exclusively, because most foundation cracks are not dormant.

• *Is strengthening across the crack required?*

If the engineer determines that it's necessary to restore tensile strength across the crack, inject an epoxy. However, restoring tensile strength across an active or moving crack may cause a new crack to appear in an adjacent area of the wall. Most urethanes have lower strengths and are not suitable for restoring strength. Select an epoxy to restore tensile strength and seal dormant cracks.

• *Is the crack wet or leaking?*

Most epoxy resins are generally not tolerant of high levels of moisture. Moisture interferes with the bond of the epoxy to the concrete, resulting in a poor seal. Select a urethane when the crack is either wet or leaking. Water activates, or catalyzes, urethanes which are even effective when water is flowing through the crack. The most common type of urethane used to seal leaking cracks creates an expanding foam upon contact with water. It foams into all crevices of the crack yielding an effective seal. Jim McCoy, McCoy Contractors, Milwaukee, says water flowing from a typical wall crack will stop immediately when injected with a urethane-based material. To repair the leaking control joint, McCoy recommends injecting a urethane for two reasons. First, the joint is leaking. Second, a urethane will not interfere with the function of the control joint. Control or contraction joints form planes of weakness permitting differential movements in the plane of the wall.

Injecting a "moving" crack

Q. *We've been hired to repair shrinkage cracks in 8-inch-thick, 8-foot-high basement walls of a 3-year-old office building. The cracks are not leaking. Yesterday we capped the cracks and were planning to inject the crack with epoxy today. But when we arrived at the jobsite, we found that the capping material had cracked. This leads us to believe that we are dealing with a moving crack. As a result, we are now having second thoughts about injecting the crack with a rigid epoxy. Should we inject the wall with a more flexible material to accomodate this movement?*

A. Your capping material probably cracked because the wall underwent small thermal movement overnight. If the walls were designed properly, they should have vertical contraction joints about every 24 feet. These joints should be able to accomodate any expected movement in the walls due to shrinkage or temperature changes.

Even if the crack in the wall was needed to accommodate movement, injecting the crack with a more flexible material, such as a polyurethane, wouldn't solve the problem. Because the thickness of the wall is far greater than the thickness of the crack, the slightest movement of the wall would have to be accompanied by an excessive amount of deformation of the injected material. The injected material would either fail, lose bond, or prevent movement of the wall.

Q. *We have to prepare random cracks in an elevated concrete slab before sealing them. Who makes power equipment that can be used for this purpose?*

A. You're probably referring to either a random crack saw or a router. Random crack saws make a ¼-inch-wide cut or wider. These walk-behind units follow the irregular path of a random crack and produce a smooth-walled sealant reservoir of uniform cross section. Dry cutting versions are available. We know of two manufacturers:

Cimline Inc.
2601 Niagara Ln.
Minneapolis, MN 55447
800-328-3874

SINCO Products Inc.
701 Middle St.
Middletown, CT 06457
800-243-6753

Crack routers use either a rotating or percussion router bit to produce a sealant reservoir. The cross section may be funnel shaped or uniform near the surface and funnel shaped at the bottom of the reservoir. Two sources that we know of are:

MacDonald Air Tool Corp.
242 West St.
South Hackensack, NJ 07606
800-328-7773

Windsor Machinery
3311 Berlin Turnpike
Newington, CT 06111
800-529-2872

Q. *We're repairing a parking garage that has several isolated cracks in the traffic deck. The cracks penetrate all the way through the slabs, and water and deicing salts leak through the cracks onto the lower levels. What is the best way to repair these cracks?*

A. We spoke to Tim Gumina at Restruction Corp. and he offered two possible solutions. The first is to inject the cracks with epoxy. He suggested injecting the cracks from the bottom up,

Crack-chasing equipment

Repairing cracks in a parking deck

leaving the tops of the cracks unsealed to get a good look at how well the epoxy is penetrating.

Another possibility is to gravity-feed either a low-viscosity epoxy or high-molecular-weight methacrylate (HMWM) into the cracks. Some low-viscosity epoxies can penetrate cracks as narrow as 0.02 inch by gravity, but it's probably better to gravity-feed only those cracks that are at least 0.05 inch wide. HMWMs, however, have very low viscosity and can penetrate hairline cracks.

To gravity-feed the materials, run a strip of caulk on each side of the crack to create a reservoir to pour the material into. When using this method, you must seal the underside of the cracks.

Seal leaks with epoxy

Q. *When injecting cracks with epoxy, occasionally the seal leaks. What can I do to continue injection after a seal leaks?*

A. If a seal leaks, try these quick repairs. Knead a 1-inch long plug of epoxy putty until it's warm. Just before placing the puttty against the leak, wipe the surface dry. Hold the putty in place for a minute or two, then start injecting again at reduced pressure.

For a leak that's draining by gravity, rub a paraffin block over the leaking or until the paraffin plugs the leak. A paraffin plug won't permit you to continue injecting, even at reduced pressures. Allow the resin already injected to set up before proceeding.

Another method that sometimes works is to put a rag over the tip of a 10-penny nail and pound it in at the leak.

Sealing below-grade walls by epoxy injection

Q. *For epoxy injection, how do you achieve surface seal on the outside of a below-grade wall?*

A. Quite often, you can't achieve a surface seal on the back side of walls. However, you can still inject the crack (about 0.01 inch or wider) with good results using an epoxy paste. Use a nonsag paste (ASTM C 881, Grade 3—nonsag consistency) that penetrates the crack but still has adequate nonsag properties so it doesn't run out of the crack's unsealed face. Choose a Grade 3 paste with the consistency of toothpaste or mayonnaise. If it's too thick, it won't enter the crack; if it's too thin, it runs out the crack and won't appear at higher ports.

Because the epoxy is a paste and not a liquid, a few variations in injection techniques are required: a) space ports as far apart as the wall is deep, b) inject at every other port, then return to inject in-between ports, and c) use higher pressures to move the paste through the pumping system and hoses or mixing heads into the crack.

Don't play with resin-catalyst ratio to change set time

Q. *We do epoxy injection work and sometimes need to slow the hardening rate of the epoxy. Can we do this by adding less hardener to the resins?*

A. No. Reducing the amount of hardener doesn't change the hardening rate of an epoxy adhesive. It simply reduces the number of chemical linkages within the hardened epoxy. Ask your epoxy supplier for a formulation that sets more slowly. You could also use an ice bath to lower the temperature of the components before mixing. Cooled epoxy, however, is more viscous and won't penetrate cracks as readily.

Chapter 6

Curing

Q. *Do you know of any data describing which curing methods produce the best results?*

A. A researcher in Puerto Rico compared the effects of several different curing methods by casting slabs, coring them, and testing the cores in compression at different ages. Core strengths were compared with the strengths of standard cylinders moist cured for 28 days. The conclusion: The most important factor affecting the strength of concrete at any time is the initial curing condition. Ponding water on the surface was found to be the best initial curing method, followed by intermittent wetting, covering with plastic sheets, and spraying with a curing compound.

Reference

R. Huyke-Luigi, "Strength of Concrete Cured Under Various Conditions in Tropical Climates," *Durable Concrete in Hot Climates*, ACI SP-139, American Concrete Institute, Farmington Hills, Mich., 1993, pp. 157-183.

Q. *Is it necessary to use a curing compound when a concrete slab is placed during the spring or fall and temperatures approach 60° F during the day but are sometimes below freezing at night? Would the curing compound possibly retain moisture that might make the concrete more susceptible to damage caused by freezing?*

A. "Cold Weather Concreting" (ACI 306R-88) advises that concrete exposed to cold weather (temperatures below 50° F) isn't likely to dry at an undesirable rate. However, since new concrete may be vulnerable to freezing when it's in a critically saturated condition, protection such as a heated enclosure may be needed. Floors are especially prone to rapid drying in a heated enclosure.

For such enclosures, ACI 306 suggests using steam to heat the concrete and prevent excessive evaporation. When dry heat is used, the report advises covering the concrete with an impervious material or a curing compound. Water curing isn't recommended since it increases the likelihood of concrete freezing in a nearly saturated condition when protection is removed.

For the conditions you mention, in which a hard freeze is unlikely, covering the concrete with insulated blankets might be the best strategy. The blankets will help to retain moisture and heat needed for curing but, when removed, will allow the concrete to dry before it's exposed to freezing temperatures.

Q. *Because shotcrete usually has a higher cement content than normal concrete, do I have to protect shotcrete from freezing as long as for normal concrete?*

Which curing method works best?

Best cold-weather curing methods

Protecting shotcrete from freezing

A. Although shotcrete typically has a higher heat of hydration because of its higher cement content, the fact that it is placed in thin layers usually offsets the heat of hydration benefits in cold weather. Therefore, protection requirements for shotcrete in cold weather are generally the same as those for normal concrete.

According to ACI 506, "Guide to Shotcrete," a minimum strength of 500 psi is required to protect shotcrete from freezing—the same as that required for normal concrete. Shotcrete placed by either the dry or wet process must be protected from freezing before the shotcrete reaches this minimum strength. Temperature during curing should be above 40° F, and ACI does not recommend using water curing in a freezing environment. Remember, also, that once protection is removed, low temperatures will prevent the shotcrete from getting stronger.

When to protect fresh concrete from freezing

Q. *The current project that I am bidding has a cold-weather clause in the specifications, which requires heating materials and protecting concrete from freezing during freezing or near-freezing weather. This specification is vague to me. What is near-freezing?*

A. Cold-weather clauses can be found in many specifications. For example ACI 318, "Building Code Requirements for Reinforced Concrete," says: "Adequate equipment shall be provided for heating concrete materials and protecting concrete during freezing or near-freezing weather."

And ACI 301, "Specifications for Structural Concrete" says: "When the mean daily outdoor temperature is less than 40° F, the temperature of the concrete shall be maintained between 50° F and 70° F for the curing period."

ACI 306, "Cold Weather Concreting," defines cold weather as a period of more than three consecutive days in which the following conditions exist:

1) The average daily air temperature is less than 40° F and

2) The air temperature is not greater than 50° F for more than one-half of any 24-hour period.

The average daily air temperature is the average of the highest and the lowest temperatures occurring during the period from midnight to midnight.

This specification may not always protect fresh concrete from freezing. For example, temperatures on a job located in a part of the country that has dramatic weather changes, like the Rocky Mountain states, may be 30° F one day and above 50° F the next. These weather conditions are not classified as "cold weather" because the temperature did not remain under 40° F for three consecutive days. Use good judgment and protect concrete whenever there is a possibility of freezing. During periods not defined as cold weather but when freezing temperatures can occur, protect concrete surfaces from freezing for the first 24 hours after placement.

ACI 306 recommendations to protect fresh concrete from freezing can be broken down into two categories:

1) Modify the mix design so the concrete will gain the necessary strength before freezing, or

2) protect the concrete from freezing by external means (i.e., blankets, enclosures, or heaters).

For more information, see "Concrete Basics" in the December 1994 issue of *Concrete Construction* on page 945 or "Cold-weather Finishing" in the November 1993 issue of *Concrete Construction*.

Quality requirements for curing water

Q. *How pure does curing water for concrete have to be? We want to use river water to cure concrete for an elevated structural slab at a train station. The river water normally contains 100 parts per million (ppm) of salt but occasionally tests at 1,000 ppm. At 100 ppm it's considered potable but at 1,000 ppm it's not. Can we safely use it to cure the concrete slab?*

A. ASTM STP 169B, *Significance of Tests and Properties of Concrete and Concrete-Making Materials*, discusses requirements for curing water. Chapter 43 says to ask two questions:
- Will the curing water stain the concrete?
- Will the curing water contain aggressive agents that might attack the concrete?

Aggressive attack is unlikely if you cure with water that's suitable for use as mixing water. Even at 1,000 ppm, the water is still suitable for mixing water according to ASTM C 94 which gives chemical limits for wash water used as mixing water.

Staining is usually caused by iron or organic matter in the curing water, not by salts. And staining from curing water is uncommon, especially where a relatively small volume of water is used.

Q. *We have just begun production on a precast bridge deck that calls for production of thousands of match cast deck segments sized approximately 3.5x7x2 meters. Our operation uses the freshly cast segment as a forming unit. We plan to accelerate the cure for the first two hours of the 12-hour curing process. Thus each segment would undergo curing twice, once as a fresh cast, then as a match piece. Does reheating precast segmental units cause any harmful effects on durability or other concrete properties?*

Match cast curing problems

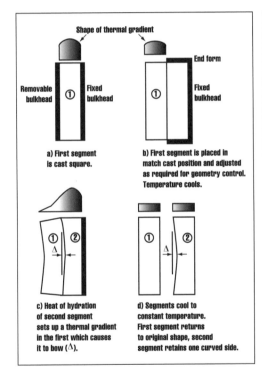

Figure 1. Bowing of match cast segments. (Ref. 1)

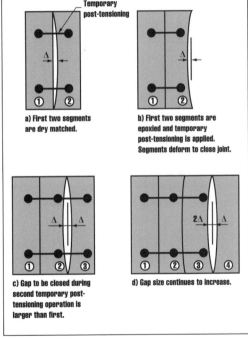

Figure 2. Plan view of temporary post-tensioning operations. (Ref. 1)

A. We know of no harmful effects on durability. In some cases, however, thermal gradients can warp the match forming unit and cause a permanent deformation in the freshly cast segment (see Figure 1). Significant deformation is likely to occur when the segment's wing-tip to wing-tip width (w) is very large compared with the segment length (L).

Warping in segments with w/L ratios of 9 or more has caused construction problems in the past. At the erection site, epoxy is applied to the matched surfaces at joints between segments prior to temporary post-tensioning that closes the gap caused by the deformation. However, the opposite side of the segment deforms because of this, thus progressively widening the gap between segments that has to be closed by post-tensioning (see Figure 2). In one case, the contractor was able to close a few consecutive joints, but then the first joint would reopen.

Researchers at the University of Texas point out that the warped segments may also result in overly thick joints and, in extreme cases, can cause cracking of the segments (Ref. 1). Their study suggested that w/L ratios exceeding 6 could present similar problems. To minimize warping, the entire match forming segment should be exposed to the same curing environment as the segment being cured (Ref. 2). However, this is often impractical since both pieces may not fit in the curing area. When both pieces don't fit in the curing area, keep the exposed part of the match forming segment as warm as possible during curing. Curing blankets and plastic sheeting may help in warmer climates. It will also help to orient the casting beds so exposure to the sun can keep the free face warm. Morning castings allow the rising ambient temperatures to warm the match forming segments as well.

References

1. Carin Roberts-Wollmann, John Breen, and Michael Kreger, "Temperature Induced Deformations in Match Cast Segments," *PCI Journal*, July/August 1995, pp. 62-71.

2. *Recommended Contract Administration Guidelines for Design and Construction of Segmental Concrete Bridges*, American Segmental Bridge Institute, March 1995, pp. 76-77.

Shrinkage cracking beyond 28 days

Q. *The specifications required 28 days curing time before applying a membrane. We applied a membrane to a deck that was more than 28 days old, and the membrane failed. We were told it was because the deck continued to crack. Is the cracking that will occur in a top deck completed in that time?*

A. Not always. The 28-day curing time is a common requirement, but don't expect all cracking to be completed by 28 days. Drying shrinkage cracking can take place months after a concrete slab is placed. The rate of drying shrinkage depends on the size and shape of the slab, the surrounding relative humidity and temperature, and the mix constituents and proportions. As an example, the Portland Cement Association's *Building Movements and Joints* indicates that a 6-inch-thick slab exposed to 70° F and 50% relative humidity will undergo only 60% of its total shrinkage by the time it's 100 days old.

Chapter 7

Defects–
Horizontal Surfaces

Q. *We made a 40-cubic-yard concrete pour for a floor on ground that was to receive a machine-troweled finish. The pour started at 7 a.m., and by 10 a.m. finishers who were edging the slab noticed a fine, white dust coming to the surface. This was occurring only along the edges; no dust appeared on the machine-troweled part of the floor. Small white dots on the surface appeared to be the source of the dust. By 5 p.m. the dust almost completely covered the area within 6 inches of the form. We've never seen this before and wonder whether it's efflorescence or has some other cause.*

The concrete contained 8 ounces of a mid-range water reducer per 100 pounds of cement, and no calcium chloride was added. I first saw the concrete at 10 a.m. and don't know how wet it was when placed.

A. We've neither seen nor heard of this phenomenon. If the concrete was very wet when placed, small channels could have been formed by excessive bleeding. Cement fines and alkalies brought to the surface through these channels may then have dried to form the deposit you describe. If power troweling sealed off the channels in the rest of the slab, fines would have appeared only at the edges. A chemical analysis of the powder would help solve the mystery.

Fine, white powder on concrete surface

Q. *We placed concrete for a basement floor with plastic hoses embedded in it for radiant heating. The concrete was placed directly on expanded-polystyrene insulation laid over plastic sheeting. A lot of bleedwater collected at the surface, so we used propane heaters to dry it up and then ran the heaters all night to help cure the concrete. The next day we applied two coats of sealer. Two weeks after the floor was placed, the owners reported that the surface was wearing away under foot traffic. They want to put a tile floor over the concrete but are concerned that the tile adhesive won't bond. What caused the problem, and what can we do to correct it?*

A. The problem is called dusting, and is caused by carbonation. If a combustion heater isn't vented, it emits exhaust gases with the heated air. The exhaust gases contain carbon dioxide, which reacts with water on the fresh concrete surface to form carbonic acid. The acid then combines with calcium hydroxide, a byproduct of cement hydration, to form weak calcium carbonates instead of the strong calcium silicate hydrates that are normally produced (Ref. 1). The soft layer caused by carbonation is usually only a small fraction of an inch deep. For reasons that aren't fully understood, dusting doesn't always occur when unvented combustion heaters are used. But to be on the safe side, you should use vented combustion heaters or electric heaters during the first 24 hours after concrete flatwork is placed (Ref. 2).

Wet grinding with a terrazzo grinder is one way to remove the weakened layer without creating dust. You can also rent grinders with a vacuum attachment that collects the dust. Remove the surface until you've reached sound concrete, and clean all remaining dust before floor tile is laid.

Cause of a dusting floor

References
1. Moira Harding, "Portable Heaters for the Jobsite," *Concrete Construction*, October 1994, pp. 800-808.
2. Steven H. Kosmatka and William C. Panarese, *Design and Control of Concrete Mixtures*, 13th ed., Portland Cement Association, 1988, rev. 1992, p. 160.

Reader response: In answer to the June 1996 Problem Clinic question in *Concrete Construction* concerning the cause of a dusting floor (pp. 510-511), you respond that the problem is due to a combustion heater's unvented gases, which react with bleedwater to form carbonic acid.

However, you fail to explain why the contractor had a lot of bleedwater on the floor surface. The excess bleedwater was due to the improper placement—directly beneath the slab—of the plastic sheeting, or vapor diffusion retarder (VDR), and the expanded polystyrene (EPS). Relocating the VDR and EPS prevents excess bleedwater and the subsequent need for heaters to dry the slab.

Gene Leger, Leger Designs, New Boston, N.H.

Editor's response: We disagree that it's improper to place EPS insulation board or a vapor retarder directly beneath a concrete slab. In our experience, freezer floors are always placed directly on the insulation board, and radiant-heated slabs (such as the slab in question) are placed directly on the insulation board or vapor retarder.

Leger recommends relocating the insulation board or vapor retarder, presumably by placing a granular blotter layer directly beneath the slab. This may cause heat to flow downward toward the cooler blotter layer instead of upward into the room to be heated.

However, we should have warned the contractor that bleeding time increases when concrete is placed directly on a nonabsorbent base. Thus, a low-slump, low-water-content concrete is needed to shorten the bleeding period.

Poor finish on parking structure slabs

Q. *I was called in to advise on problems with a parking structure. The owner is unhappy with the finishing work on all four slab levels. He is refusing to accept nearly 400,000 square feet of surfaces with trowel and bull float marks, footprints, and unclosed ragged rough areas where the concrete was not properly consolidated. Since preliminary proposals for correcting the situation show costs around $6 per square foot, we're looking for alternative ways to fix it.*

Fortunately the job is in a southern state where freeze-thaw durability is not a concern. The slab concrete meets strength specifications, testing 4000 psi or better. There has been no auto traffic on the slabs yet.

A. A thin bonded overlay, such as those used for bridge deck repair, is one possible remedy. Probably the thinnest, and among the most costly, would be a polymer concrete made with resin and aggregates, using no portland cement. These overlays are installed from ¼ to ⅜ inch thick. Latex-modified concrete would probably cost less. It can be installed in a bonded overlay as thin as 1¼ inches, although a minimum of 1½ inches is recommended (see *Concrete Construction*, March 1987, page 261).

Another overlay alternative is the so-called Iowa low-slump concrete, which can be installed 1¼ inches thick. This mix, used on more than 1,000 bridge decks in Iowa, contains 823 pounds of cement per cubic yard and is placed at a ¾-inch slump. A mix this stiff requires special consolidating equipment, and some states have been trying the same mix with superplasticizers added to ease placing problems (see *Concrete Construction*, February 1988, page 131).

Since the parking structure is in a mild weather zone, something less than these bridge deck overlays would be adequate. A thin conventional concrete topping ¾ to 1 inch thick could probably be installed using a good epoxy bonding agent. Ross Martin of Baker Concrete Construction suggested another economical solution—mechanically texturing or grinding the surface, removing about ½ inch of depth to get a more satisfactory, uniform appearance. But, he cautioned, you must be sure that this doesn't reduce cover of the reinforcing steel below specified limits.

Consider also the possibility of placing well-bonded patches in the distressed areas, then installing a traffic-bearing membrane like those described in *Concrete Construction*, June 1990 (page 545). These two-layer systems form a tough, impervious coating. Some are flexible enough to bridge small cracks, and they're available in a variety of colors.

Q. *I recently looked at a discolored driveway that has remained discolored for more than a year after placement. The finisher used a bull float, then a steel trowel, and finally a broom. The granular base beneath this part of the driveway was used as the truck washout area when the upper part of the driveway was poured. Could this be related to the discoloration?*

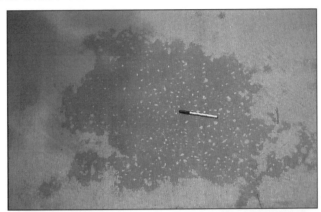

A. *Reader response:* I have seen many examples of discoloration and some with a startling contrast between the light and dark sections. So many such problems have occurred in the lower mainland of British Columbia and on Vancouver Island that a very descriptive name has been given to the problem. It is referred to as "pinto" concrete because it resembles the coloration of a pinto horse.

So far as I'm aware, nobody has a clear picture of what causes pinto concrete and I am equally unaware of a cure for the condition once it has occurred. I can make some general statements from my own observations but I am aware of some exceptions to these generalizations:

1. Pinto is a problem confined almost exclusively to residential driveways and sidewalks. I have not personally seen it on city sidewalks which generally are of higher-strength concrete and are not troweled. Nor have I observed it on interior floors.

2. Pinto is not a widespread problem in our area but is more or less confined to the greater Vancouver area and parts of Vancouver Island.

3. Cases of pinto are reported in bunches with periods of several years passing in which no reports reach my attention.

4. The dark and the normal-colored concrete generally come from within the same load of plastic concrete.

5. I have never been able to explain the color difference by any differences in the underlying subgrade.

6. Although the contrast can become less pronounced with age, pinto is usually a permanent condition. I have observed it in concrete that is 10 years old. I also recall one extremely bad case which completely cleared up with the advent of warmer weather.

The most troubling and intriguing thing about the phenomenon is the sharp demarcation between light and dark in a product that is subject to as much mixing and blending as is ready-mixed concrete.

Paul M. Heaton, Tilbury Cement Ltd., P. O. Box 950, Delta, BC, Canada V4K 3S6

The discolored driveway appears to be blotchy as a result of using fly ash. This causes pinto color. To prevent this problem, specify to your ready mix supplier that there is to be no fly ash used in your mix.

Andrew J. Pucciarelli, Pucciarelli Bros. Concrete Contractor, 103 S. Harleston Dr., Pittsburgh, PA 15237

We have experienced the same type of problem especially over a wet subgrade, which apparently affects hydration of the ferrites. We have been successful in reducing the discoloration by applying pure vinegar and allowing it to stand on the surface for about 30 minutes before brooming it off using lots of water. This opens the surface and allows hydration.

In some cases a second application has been necessary. But there has always been a marked improvement, if not total elimination, of the problem. We have always recommended using a sealer following the above procedures.

Stanley Ernst, Ernst Concrete & Supply Co., 23055 Croesbeck Hwy., Warren, MI 48089

The question of what causes discoloration was studied in detail at the Portland Cement Association during the mid-1960s. I started the study, was soon joined by Nate Greening and was subsequently replaced by Robert Landgren after I left for Dundee Cement Co. The study is reported in the PCA Research Dept. Bulletin 203, "Surface Discoloration of Concrete Flatwork."

Discoloration of the type shown, and several other types, was produced at will. All that was required was to use calcium chloride in the concrete, apply a hard-trowel finish, and neglect moist curing for the first day. The discoloration was caused mainly by retardation of the tetracalcium alumino ferrite phase of the cement by the chloride. Tetracalcium alumino ferrite is the darkest compound usually found in cement. When it is retarded it retains its dark color; hydration lightens it markedly. Hard troweling increases the density of the concrete surface and therefore darkens it.

Remedial treatments found to be effective were immediate washing with water or with sodium hydroxide solutions. The most effective was the application of a 20% to 30% solution of diammonium citrate.

The best ways to avoid the problem were to avoid the use of calcium chloride; use positive curing, including excess water for at least the first day; and avoid hard troweling (which isn't necessary or even desirable on a driveway slab).

William E Perenchio, Wiss, Janney, Elstner Associates Inc., 330 Pfingsten Rd., Northbrook, IL 60062

As service engineer for a ready-mixed concrete company, I had to answer a complaint similar to the one you describe. Our customer was doing patios and service walks in a new home building project.

The excavator had spread the topsoil for the lawns all around the homes, but our customer was not removing this dry black dirt when he graded the areas to be covered with concrete. In some places he had black dirt showing, and in others he might have had 4 to 6 inches of pea gravel. The complaint was about discolored concrete just like that shown in your photos.

I think the cement was hydrating at a different rate in the areas where dry black dirt was absorbing water from the concrete faster than moist pea gravel does. In the problem you describe, differing rates of subgrade absorption could be caused by the subgrade being wet in spots.

Joe Seyl, Meyer Material Co., P.O. Box 129, Des Plaines, IL 60016

I have been continuously involved in the concrete industry for the past 20 years as a laborer, carpenter, finisher, contractor, and inspector. I'm currently a public works inspector in California but also have experience in the Chicago area.

I have seen many examples of the discoloration shown in your photos. In one case my client was so displeased with the discoloration effect that I demolished the concrete in and around the affected spot. I found a locally concentrated area of clay subgrade. During preplacement preparation, subgrade materials were removed to varying depths. This left scattered spots of thick clay and other spots where granular material was used as fill.

Clay retains moisture for long periods of time when unexposed to sun and evaporative winds. This means there were different concrete curing conditions due to variable available moisture in the subgrade. This is consistent with the reader's description of his conditions where parts of the subgrade were used for truck washout.

I have seen the condition prevail for years before the water actually evaporated or was consumed by the extended hydration process and the paving finally bleached out to match the adjacent surfaces.

Portions of concrete flatwork that have been left until "dead" (after initial set) before hard troweling also have exhibited similar discoloration. I believe this is caused by differences in density. "Burning in" a very thin layer of fine sand and cement particles causes formation of very hard, often brittle, areas within that section of the slab.

Attention to subgrade preparation and to concrete placing and finishing techniques alleviates discoloration problems.

James R. Thomson, Priority Inspections, 2210 Dalecrest Lane, Spring Valley, CA 91977

I have been in business as a concrete contractor for the past 42 years. I have seen this mottling discoloration many times. Many years ago, together with the Portland Cement Association and Wagnald Ready Mix Co., we did some investigative work regarding this problem.

We concluded that this dark blotching is due to the subgrade being extremely wet or saturated under the blotched areas. This, we found out, will not occur if a sand-type base (2 inches thick or more) is used under the grade slabs. In fact, a sand-type base should always be used under grade slabs that are colored, whether integral or applied as a topping, to help assure a uniform color.

The only way we discovered to eliminate this dark blotching discoloration after it occurs is by using a hand-type blow torch and applying the flame directly to the concrete surface. This will remove most of the darkness and help produce a more uniform, light appearance. Doing this is time-consuming and it may not be practical for large areas of discoloration.

Frank P. Petrilli, Frank P. Petrilli & Son Inc., 20654 Bahama St., Chatsworth, CA 91311

Your reprint collection, *Troubleshooting Concrete Flatwork and Paving Problems*, gives some of the possible causes for discoloration. It also recommends using phosphoric acid for treating the discoloration. I have found that white vinegar works as well as phosphoric acid and is much easier to find. Red vinegar sometimes leaves a red tint. This usually disappears but causes questions until it goes away.

Keeping the trowel off outside concrete will eliminate 90% of discoloration problems and also will eliminate a large percentage of the scaling problems caused by freezing and thawing. Water is trapped under the densified layer caused by troweling.

James E. Lunsford, Signal Mountain Cement Co., 1201 Suck Creek Rd., Chattanooga, TN 37405

In my 30 years of looking at concrete problems, I have seen this problem many times on driveways. The finishers had always used a steel trowel before brooming. Steel troweling air-entrained concrete brings too much water to the surface and this requires additional steel troweling so that late troweling will occur in areas of the driveway. Late troweling always causes dark areas. Brooming will not cover this up.

If the finisher had used a magnesium float instead of the steel trowel, discoloration wouldn't have been as likely.

Discoloration is also more likely when concrete is placed at a 5- to 8-inch slump as is done with many driveways.

I recommend placing concrete at a 3- to 4-inch slump, bull floating it, then brooming it. It also must be effectively cured. If the finisher had used a white-pigmented curing compound, the discoloration would have been covered until after the first winter and then wouldn't have been as noticeable. That's because white pigment in the pores will lighten the dark areas.

Roy B. Worstall, Worstall Concrete Consulting Co., 2023 Marion Ave., Zanesville, OH 43701

Very little information was given with regard to ambient conditions at the time of placement, curing procedures, mix design, and slump at the time of placement. Because of this, I can only provide the following possible causes and remedies.

Possible causes

1. Changing cement brands or fine and coarse aggregate in consecutive batches of concrete placed monolithically in the forms.

2. Wide variations in concrete mixing water causing varying amounts of laitance to appear at the surface.

3. Hastening the finishing operation by dusting cement onto the wet surface to remove bleed water.

4. Discontinuity in the steel troweling operation resulting in a varying degree of surface densification.

5. Nonuniform application of curing compound or uneven coverage of the concrete surface by wet burlap or polyethylene.

Possible remedies

1. Avoid changes in concrete ingredients and proportions from batch to batch for the same class of work.

2. Ensure uniformity in placing and finishing for concrete placed monolithically.

3. Avoid using dusting agents such as cement or calcium chloride to expedite the finishing operation by drying a water-laden surface.

4. Provide adequate and uniform curing to exposed slab surfaces.

Jeffry S. Stoker, Master Builders Inc., 10999 Reed Hartman Hwy., #105, Cincinnati, OH 45242

I suggest you review PCA research on discolored concrete flatwork to shed some light on discoloration in concrete. Some of my colleagues suggest the lack of or slowed hydration of ferrites as a probable cause.

I would also like to comment on fly ashes. Fly ashes come in many different colors—white, buff, beige, light brown, limestone (light) gray, medium gray (between the color of Type I and Type II cement), and dark gray and brown. Fly ash, if used at about 100 pounds per cubic yard (which is somewhat normal for northern climates), and if light in color, will have negligible color effect.

However, darker fly ashes will uniformly tint concrete either gray or brown. If concrete is dramatically darkened, you may question the producer as to the quality of the fly ash used, a possible cement change, or excess minus No. 200 material present in fine or coarse aggregate. Experience has shown that if aggregates have a 4% or 5% wash, this can dramatically tint concrete brown or gray depending on aggregate types.

In defense of fly ash in eastern Pennsylvania, my company over the past six or seven years has provided a good quality, light-colored fly ash represented by about 10 million cubic yards of concrete. This concrete represents all types of structures and pavements.

Richard A. Mackow, JTM Industries Inc., Allentown, PA

I am a supplier of both class F and C fly ashes and have worked with it exclusively in Wisconsin, Illinois, Indiana, and Michigan for the past 9 years. In this time I have never been involved in or heard of an instance where fly ash was the culprit of blotchy discoloration.

I am very surprised that *Concrete Construction* would print such a response when it is apparent that the discoloration was caused by both a subgrade condition as well as improper finishing.

James L. Hegle, National Minerals Corp., Pleasant Prairie, WI

The responses to "Causes and Solutions for Slab Discoloration" were quite informative. I was impressed with the willingness of persons in our industry to share their expertise.

On the other hand, I was very distressed to read the response saying, "The discolored driveway appears to be blotchy as a result of using fly ash." This technically unsubstantiated statement offends the responsible use and user of good quality fly ash. The implication that whenever one uses fly ash, blotchiness or pinto color will appear is certainly not the case.

Daniel Stevens, American Fly Ash Co., Naperville, Ill.

Editor's postscript: After publishing the Problem Clinic item, we got more information from the concrete supplier. The air-entrained concrete contained no calcium chloride or fly ash. It was in a 5-yard load with a 12-mile haul in a truck with fins in good shape. Thus, undermixing is unlikely. Slump was 3 inches and the contractor placed the concrete at that slump because the driveway was sloped.

Can footprints in concrete be removed?

Q. *While applying curing compound, one of my workers walked on a parking structure deck too soon. He left footprints in the concrete that are very shallow, less than 1/16 inch deep, but still visible. There are many of these footprints in areas of up to 900 square feet, and the owner wants them repaired so they aren't noticeable. Is there any way to do this economically?*

A. It might be possible to remove the footprints by shotblasting the areas containing them. Although this will probably change the color and texture of the treated areas, the overall appearance might be more acceptable to the owner. Production rates for large shotblasting units can exceed 3,000 square feet per hour. Consult a contractor who does surface preparation for floor coatings. You can find a list of some of these contractors in the membership directory of the International Concrete Repair Institute (ICRI). For information about the directory, call ICRI at 703-450-0116.

You could also try patching the footprints, but this will make them more noticeable. It's also difficult, if not impossible, to ensure bond for such a thin repair.

A third, more expensive alternative is to patch the footprints and then apply a thin coating or membrane that will cover the entire affected area. This also will change the color and texture of the treated area and probably require the services of a skilled contractor to apply the coating.

The footprints don't impair the function of the structure, and repair options are likely to leave something just as conspicuous as the footprints. It's also possible that the shotblasting option will open the surface, allowing chlorides to more easily penetrate the concrete. In view of these potential disadvantages, we'd suggest offering the owner a credit in lieu of trying to obliterate the footprints.

Circular cracks probably caused by overloading

Q. We moved into a 60,000-square-foot tilt-up concrete building six months ago. The building is two years old but we're the first tenant. The floor is 5 inches thick and contains 6x6—W2.9x2.9 welded wire mesh. Control joints are 20 to 25 feet apart. Before we bought the building, we had an environmental audit done. As part of the testing, three cores were taken from the floor and tested in compression. The average core strength was more than 9000 psi.

Since we occupied the building, we've noticed nine circular or football-shaped cracks in the floor. They're randomly located but are generally near the building perimeter. Diameter of the circles ranges from 9 feet to 16 feet. In some of the circles there are also radial cracks. Some of the circular cracks are in the middle of a bay, away from any joints, and some intersect sawed joints (see photo).

Because the building isn't very old and hasn't been in use long, we're concerned that the cracking will get worse. Are we experiencing subgrade settlement, shrinkage cracking, or something else? What testing can be done to determine the cause of these cracks?

Some circular floor cracks occurred at midpanel and some intersected joints. Cracks are painted yellow. Overloading is a likely cause of the cracking.

A. The cracks don't look like shrinkage cracks and are most likely to have been caused by a concentrated load. When there's a concentrated load on a slab, the bottom of the slab beneath the load is in tension and radial cracking results. At some radial distance from the load, the top of the slab is in tension and this causes a circular cracking pattern.

Since the structure is a tilt-up building, overloading may have occurred under the crane outriggers when panels were being lifted. A similar pattern could show up, however, if the floor is thin in spots and is subjected to heavy concentrated loads under service conditions. Taking cores at some of the cracked areas would allow you to check floor thickness.

Circular cracks around columns

Q. *We built a slab on grade in a building with rectangular concrete columns. We isolated the columns from the slab by wrapping them with a compressible material. Roughly circular floor cracks developed around some of the columns. Can shrinkage cause this kind of cracking, or is there another more likely cause?*

A. Circular cracks in floors are often caused by concentrated loads. At some radial distance from the load, the top of the slab is in tension, and this causes the circular cracking pattern. Such loads can be caused by settlement of column footings if the columns aren't free to move independently of the floor slab. For a thorough diagnosis, a consulting engineer should examine the floor, column isolation joints, and cracking pattern.

D-cracking vs. ASR-cracking

Q. *We've got a series of parallel cracks along the joints in our concrete parking lot. An engineer says they're D-cracks. Is this kind of cracking caused by alkali-silica reaction (ASR)?*

A. No. D-cracking is caused by using coarse aggregates that are susceptible to freezing-and-thawing deterioration. The aggregates absorb moisture from the pavement base and from surface water entering through cracks and joints. If aggregate pores are full when freezing occurs, internal pore pressure cracks the particles, causing the mortar to crack as well. More cracks develop with repeated freeze-thaw cycles. D-cracks are roughly parallel to the adjacent joint. Many individual cracks caused by alkali-silica reaction are approximately perpendicular to the direction of the joint. These ASR cracks are also commonly associated with fainter map cracking elsewhere in the pavement slab.

Handbook for the Identification of Alkali-Silica Reactivity in Highway Structures, by David Stark, is an excellent aid to detecting ASR damage and distinguishing it from other types of concrete damage. Published by the Strategic Highway Research Program, the 49-page, color-illustrated handbook can be purchased from the Portland Cement Association. Call 800-868-6733 and request publication LT165.

Avoiding scaling and sheet scaling

Q. *I'm the construction coordinator for a city in a freeze-thaw region. Deicer scaling and sheet scaling of city streets are both a concern to us because either form of this distress can affect pavement aesthetics and cause surface roughness. What are some of the causes and ways to prevent this kind of deterioration?*

A. Although both problems are called scaling, the causes are different. Deicer scaling commonly occurs on exterior flatwork that's exposed to freeze-thaw cycles and applications of deicing agents. The finished surface flakes or peels off, usually in small patches that may later merge to expose larger areas.

During freezing and thawing, hydraulic pressures that develop within the concrete may be sufficient to cause scaling. Deicing agents accelerate scaling, probably because applying the agents increases the number of freeze-thaw cycles and helps keep the concrete moisture content above a critical level at which scaling occurs. Deicers also dissolve in the concrete pore water, creating osmotic pressures that add to hydraulic pressures.

To avoid deicer scaling:

- Use air-entrained concrete (5% to 8% air content).
- Limit concrete slump to 5 inches.
- Don't perform any finishing operations when water is on the concrete surface.
- Cure the concrete with a curing compound or watertight covering, then allow it to dry uncovered for at least 30 days before deicing salts are used.

Even if exterior flatwork isn't exposed to freeze-thaw cycles or deicers, it can exhibit sheet scaling. The surface peels off in areas ranging in size from several square inches to many square feet. Sheet scaling occurs after a layer of water and air collects beneath a dense-troweled surface. Troweling seals the surface while the underlying concrete is still bleeding or able to release air. Because air entrainment slows bleeding, finishers are more likely to trowel too soon, especially on hot, dry, windy days. Placing concrete on a cold surface also contributes to the problem because the concrete sets more slowly and thus bleeds longer. To avoid sheet scaling:

- Prohibit steel troweling of air-entrained concrete. Instead, call for a float, broom, or burlap-drag finish.
- Use heated or accelerated concrete when placing concrete on a cold surface.

Excellent summaries of scaling and sheet scaling are given in the Concrete in Practice (CIP) series, published by the National Ready Mixed Concrete Association, 900 Spring Street, Silver Spring, MD 20910 (301-587-1400). Ask for CIP 2, *Scaling Concrete Surfaces*, and CIP 20, *Delaminations of Troweled Concrete Surfaces*.

Causes for slab discoloration

Q. *I'm a perfectionist when it comes to concrete finishing. I want to know why discoloration occurs in some instances on recently placed slabs. The discolorations are usually light and dark patches. I know that concrete slabs eventually bleach out with exposure to the sun, but sometimes the building shell is built before bleaching occurs, leaving a permanently discolored slab. I've asked many finishers why this discoloration occurs. Their answers have included spraying water on the surface to make finishing easier, excessive finishing, and improper placement of thermal blankets. Are there other possible causes?*

A. We asked our readers this same question in June 1991 and published many suggested causes in *Concrete Construction* in October 1991. These included steel troweling, variable subgrade composition, and variations in the concrete's water-cement ratio. One reader cited a detailed study conducted by the Portland Cement Association (PCA) in the mid-1960s. This study showed that discoloration could be produced by using calcium chloride in the concrete, applying a hard-trowel finish, and failing to moist cure for the first day after concrete placement. Discoloration was caused primarily by calcium chloride retarding the reaction of tetracalcium aluminoferrites in the cement. When hydration of this compound is retarded, it retains its dark color; hydration lightens it markedly.

Some discoloration that's reported occurs when there's no calcium chloride in the concrete and no hard troweling. This form of discoloration is believed by some to be related to water that's trapped just beneath the surface during finishing. It seems to occur most often in coastal areas with high ambient relative humidities.

For further discussion of the problem and some solutions, see "Pinto Concrete: Is There a Cure?" in *Concrete Technology Today*, Vol. 17, No. 1, March 1996, p. 4. You can purchase a copy by calling PCA at 800-868-6733.

Q. *We have footprints on a floor that has a hard-troweled finish with a black dry shake hardener. The floor was sealed with a high solids acrylic cure and seal product. We've used a steel wool buffer to try to remove the footprints, which are in an approximately 1,200-square-foot area. Are there any other methods we could try?*

A. First determine whether the footprints are in the sealer or in the floor surface. In a small area with one of the footprints, remove the sealer with a strong solvent such as a paint stripper, xylene, or toluene. Ask the sealer manufacturer for a recommended solvent and be sure to take all needed safety precautions when using the solvent. If the footprints come off with the sealer, use the solvent on all affected areas. If the footprints are still visible after the sealer has been removed, light grinding will remove them. However, grinding will expose fine aggregate in the dry shake and will probably change the floor color. You might make the situation worse instead of better.

Removing footprints on a floor

Q. *We supply concrete for two large flatwork contractors who each place several hundred thousand square feet of concrete floors per year. One experiences curling on some jobs but the other doesn't. Both use 4000-psi concrete and do a good job of curing.*

There's one difference between the contractors. One uses concrete with 1¼-inch-maximum size rock for all floors 5 inches thick or greater. He doesn't get any curling. The other uses ¾-inch-rock concrete regardless of floor thickness. His floors sometimes curl.

An engineer told me he thinks the large-rock concrete doesn't curl as much because of better aggregate interlock. Is that correct? Or are there other factors involved?

A. The aggregate interlock theory may explain what you've noticed, but there's another factor that may be more important.

Concrete curls when the top of the slab shrinks more than the bottom does. The more a slab shrinks, the more it's likely to curl. So reducing shrinkage reduces curling as well.

Concrete made with larger rock doesn't require as much water per cubic yard to produce the same slump as a smaller-rock concrete. The lower water content reduces shrinkage and also should reduce curling.

Bigger aggregate reduces curling

Q. *We have to troubleshoot exterior flatwork problems including scaling. Is there any way to tell if scaling was caused by deicing salts and freeze-thaw cycles or poor finishing methods?*

A. An inadequate air void system can cause scaling of concrete exposed to freezing and thawing and deicers even if the concrete is finished correctly. Having a petrographer check air void characteristics of the hardened concrete will tell you if there's a problem with the air void system.

Some scaling is caused by poor finishing methods and may occur even if the concrete is properly air entrained or isn't in a freezing climate. Called sheet scaling or delamination, it can be traced back to closing the concrete surface by troweling before bleeding has ceased. Bleedwater causes a weakened plane when it collects under the thin layer that's densified by premature finishing.

How to tell if scaling is caused by deicing salts or finishing problems

Examine pieces of the mortar that have scaled off. Petrographer Mauro Scali says that if the scaling is caused by premature finishing, the top of the piece will generally be darker and the bottom of the piece will be lighter in color. When scaling is caused by low air content, the top and bottom of the pieces are usually the same shade of color.

Correcting a wet garage floor

Q. *I live in Louisiana. Many times my garage and shop floors become very wet and slippery during the winter, when the air temperature changes quickly from cold to warm. What can be done to concrete floors to prevent this from happening? I'm also installing a new shop floor and would like to know what can be done during construction to prevent this problem.*

A. Water is probably condensing on the floor surface due to temperature differences between the air and the concrete. Warm air holds more moisture than cold air. When the air temperature increases rapidly, the concrete temporarily remains colder than the air, causing moisture in the air to condense on the floor. To verify this, use duct tape to seal the entire perimeter of a 2x2-foot square of plastic film to the dry floor. When the floor gets wet again, check to see if it's wet under the plastic. If it isn't, condensation is the problem.

If your shop or garage area is enclosed, a dehumidifier would help keep the floors dry. You might also consider grooving or grinding the floor to reduce the slipping hazard. You can rent grinders, groovers or scabblers from equipment rental houses. The disadvantage of rough-textured surfaces is that they are hard to keep clean.

For new construction, consider placing the floor on 2-inch-thick insulation board. Installing in-slab radiant heating will decrease the temperature difference between the floor and air, and possibly prevent condensation, but this is a more expensive alternative.

Blistering in heated enclosure

Q. *Would a natural gas heater 3 feet above the floor cause blistering? We supplied concrete for a garage floor, and the flatwork contractor had trouble finishing it in the vicinity of the heater. He said concrete near the heater felt like bread dough during finishing, and afterward he found the blisters in that area. The area was only about 2 feet square.*

A. It's likely that the blisters were caused by the gas heater. Moderately rapid evaporation of bleed water makes the surface ready to be troweled while the underlying concrete is still bleeding or still plastic and releasing air. Troweling seals the surface so that bleed water or air from below can't get through. Blisters form sometime after the first troweling. Cold concrete or a cold subgrade would aggravate the problem. Low temperatures retard setting and increase the bleeding period of concrete beneath the warmed surface.

Air-entrained concrete with a high cement content or made with fine sand is most likely to blister. Thick slabs and slabs placed on a polyethylene vapor barrier also are more likely to blister.

There's an excellent discussion of blistering available from the National Ready Mixed Concrete Association. It's called *Concrete Blisters: What, Why, and How.* To order copies write to NRMCA, 900 Spring Street, Silver Spring, MD 20910.

Can dusting floors be cured?

Q. *I know that dusting—the development of a powdered material at the surface of concrete—is associated with weak concrete at the surface. I've also learned several ways to prevent dusting during new construction. However, besides grinding the surface, I've never heard of any methods to correct dusting once it has occurred. Are there any?*

A. Dusting can be remedied by applying a floor-sealing product based on sodium metasilicate (water glass) or silicofluorides.

When a dilute solution of sodium metasilicate soaks into a floor surface, the silicate reacts with calcium compounds to form a hard, glassy substance within the pores of the concrete. The degree of improvement depends on how deeply the silicate solution penetrates. For this reason, dilute the solution with water to make it penetrate deeply.

Apply three or four coats, allowing the previous coat to dry before applying the next coat. For the first two coats, use four parts water to one part silicate. The third coat should be a three-to-one solution applied after the second coat has dried. The treatment is completed as soon as the concrete surface gains a glossy, reflective finish.

Zinc, sodium, and magnesium silicofluoride sealers are applied in the same manner as water glass. These compounds can be used individually or in combination, but a mixture of 20% zinc and 80% magnesium gives excellent results. For the first application, dissolve 1 pound of silicofluoride in 1 gallon of water. For subsequent coatings, use 2 pounds per gallon of water. Mop the floor with clean water shortly after the preceding coat has dried to remove encrusted salts. Carefully observe all safety precautions when applying silicofluorides because they are toxic.

It is important to note that sealing products will not convert a poor-quality floor into a good-quality floor. They simply are a means of upgrading a dusting floor while improving its wear resistance.

Reference

ACI 332, "Guide to Residential Cast-in-Place ConcreteConstruction," ACI 332R-84.

Q. *We get a lot of popouts in exposed concrete. The expansive materials are shale particles found in the sand. Even though the shale particles may be no larger than ⅛ inch in diameter, a conical fracture shape makes the popouts more noticeable. Can this problem be avoided?*

Preventing popout problems

A. The popouts you describe occur primarily in Iowa and Minnesota. Shale particle expansion results from a reaction between the shale and alkalis in the cement, not freezing. The reaction occurs too fast for pozzolans to help.

The slab in the top photo was finished and then sealed with a normal sprayed sealer. The slab in the bottom photo was finished the same way but cured using wet burlap bags. Its concrete is slightly discolored from the bags.

The Des Moines Metro Concrete Council placed test slabs to evaluate methods for reducing popouts. Most successful was a wet burlap cure, believed to leach out alkalis near the surface so they won't react with the shale. Slag cement instead of straight portland cement also reduced popouts. Use of a manufactured stone sand eliminated popouts but may have reduced the concrete's abrasion resistance.

For information, contact Byron Marks, Martin Marietta Aggregates, 4554 N.W. 114th St., Des Moines, IA 50322.

Popouts and low-chert aggregates

Q. *We are a ready-mixed concrete producer in the Midwest. Much of our business comes from supplying concrete for residential and commercial patios, sidewalks, and driveways. We sometimes receive complaints about popouts occurring in these concrete surfaces. We don't think it is the fault of our concrete mix, and the concrete contractors we work with seem to be finishing the concrete properly. What causes popouts and is there any way to prevent them?*

A. Popouts are indentations in the surface of concrete left when aggregate particles (stone or sand) have expanded and worked themselves loose. These particles are termed "deleterious" by ASTM. A dictionary definition of deleterious is "harmful, often in an unexpected or subtle way." These deleterious particles most likely separate from the concrete as a result of either physical or chemical reactions.

A *physical expansion popout* can occur if lightweight, porous, deleterious aggregate takes on moisture, freezes, expands, and fractures. Often this fractured piece of aggregate can completely separate from the concrete, taking a portion of the surface mortar with it. The resulting indentation in the surface of the concrete is called a popout.

A *chemical reaction popout* occurs when alkalies in the portland cement in the concrete react chemically with silica present in some fine sands, causing an expansion of the silica particles, resulting in small surface indentations.

Unless they are severe in size and number, popouts do not decrease the service life of a concrete slab. However, some people do find popouts aesthetically unpleasant.

Some parts of the country, such as the Midwest, have a variety of soils, sand, and rock. This was caused by glaciers during the last Ice Age depositing these materials as they receded. Many of these deposits contain small percentages of deleterious materials, including certain shales, iron oxides, unsound cherts, and other soft particles.

Ready-mixed concrete producers most often use locally available sand and rock deposits to make concrete. Sometimes, these local deposits contain small amounts of deleterious materials, including some unsound cherts, that can cause popouts.

Local building codes, the American Concrete Institute, and ASTM construction specifications recognize that it is impractical to completely eliminate all deleterious materials from aggregates used to make concrete. By allowing for a small percentage of deleterious materials to be used in codes and specifications, these agencies and associations accept that some popouts may occur.

Concrete surfaces with popouts can be repaired. Small patches can be made by cleaning out the indentation caused by the deleterious particle and filling the void with dry-pack mortar, epoxy mortar, or other patching materials. When popouts are too numerous to patch individually, a thin bonded concrete overlay can be placed over the existing concrete surface.

To minimize popouts, consumers can specify performance concrete mixes for exterior flatwork that include:
- Low-chert aggregates
- Cements meeting ASTM specifications for alkali-aggregate reactivity
- Air entrainment of 6% ± 1%
- Moderate slump of 5 inches ± 1 inch
- Strength of 4000 psi in 28 days.

Reader response: As chairman of the Technical Committee of our state association, Aggregate Ready Mix of Minnesota, I deal with many of these same problems at both the state and company levels. One of my goals the last several years has been to produce simple, accurate, and technically correct brochures for producers to use when dealing with these problems.

One brochure deals with popouts. One item the brochure stresses is acceptance of popouts due to use of local aggregates. A chart in the brochure, shown below, even projects the potential number of popouts that may occur.

Limits of deleterious substances of aggregate for concrete

Coarse particles	Coarse aggregate			Fine particles	Fine aggregate	
	ASTM* C33-90 Class 4S	MNDOT #3137 General use	MNDOT #3137 Bridge super-structure		ASTM* C33-86 Class 4S	MNDOT #3136 General use
	Maximum allowable				Maximum allowable	
1. Clay lumps and friable particles	3.0%	3.5%	3.0%	1. Clay lumps and friable particle	3.0%	—
2. Chert (less than 2.4 SpGr SSD) total spall (MnDot)	5.0%	1.0%	0.3%	2. Coal and lignite	0.5%	0.3%
3. Combined 1 and 2	3.0%	3.5%	3.0%	3. Other deleterious substances (shale, mica, soft and flaky particles)	—	2.5%
4. Coal and lignite	0.5%	—	—			
5. Shale On ½-inch sieve Total Sample	— —	0.4% 0.7%	0.27% 0.3%			
6. Soft iron oxide	—	0.3%	0.2%			
Potential density of popouts per sq. yd. with maximum deleterious limits**	20 to 30	15 to 20	5 to 10			

Note: The most troublesome deleterious particle in the Midwest, shale, is not identified in the national ASTM specification.

* S Category, severe weathering region: Areas with greater than 500 day-inches weathering index (product of the average annual number of freezing cycle days and the average annual winter rainfall in inches). Meets uniform building codes for commercial and residential construction.

** Due to unusual circumstances or weather conditions during placing, finishing, or curing, the number of popouts may vary from those predicted.

Source: Aggregate Ready Mix of Minnesota

One of the greatest benefits of these tools is that they are backed by an association and a committee made up of members involved in various aspects of our industry. It is sometimes difficult for a producer to tell a contractor, whose business is placing concrete, that he's doing it wrong. But when you have an association behind you and a panel of experienced members, backed up with information from ASTM, ACI, PCA, etc., your case for correct procedures and techniques is much more solid.

Jerry E. Lang, Operations Manager, REMIX Concrete Co., Milaca, Minn.

Although the information in your July section on popouts was not new, the presentation was clear and concise. This could have been the perfect presentation to a magistrate in defense of "defective concrete resulting from a few popouts."

Unfortunately, the last paragraph advises the consumer to order concrete with low-chert aggregate. This assumes that we have two separate stockpiles: high-chert and low-chert. The recommendation is not a solution. It simply confuses the issue. The plant troubleshooter would have been better served by simply recommending that aggregate conform to ASTM C 33.

Robert L. Yeske, President, 43rd Street Concrete & Asphalt Co., Pittsburgh

Author's reply: Mr. Yeske is correct in suggesting that ASTM C 33-90 be referenced when defending "defective concrete resulting from a few popouts." Indeed, my article mentioned that ASTM, ACI, and many state and local building codes recognize it is impractical to completely eliminate all deleterious material from concrete aggregate and thus accept that some popouts may occur.

But in southeastern Wisconsin, consumers are willing to pay a premium price—between 12% and 25%, depending on the low-chert aggregate source—to reduce concrete popouts in driveways and patios. With a willing market, southeastern Wisconsin ready-mixed concrete producers are glad to value-add residential exterior concrete flatwork, and most do keep two aggregate stockpiles: one conforming to ASTM C 33-90 and a separate stockpile of low-chert aggregates.

Jerry Krueger

Is craze cracking a terminal illness?

Q. *We just completed a concrete floor that is acceptable in every respect except one: There are numerous craze cracks over much of the surface. The floor flatness meets specification requirements, and there are few random shrinkage cracks. However, the owner is concerned that the craze cracks will cause progressive floor deterioration. Is there any information in the literature to show that floors with craze cracking can still be durable?*

A. Crazing is generally believed to be a cosmetic problem. The cracks are unsightly, but seldom affect surface durability or wear resistance.

There are some exceptions, however. We know of one laboratory study in which crazing was purposely induced by applying a dry cement shake to a nearly two-hour-old slab with bleedwater on the surface (Ref. 1). After the shake had absorbed bleedwater for five minutes, an experienced finisher steel troweled the slab surface. This surface and others in the experiment were exposed to rolling friction, and depth of wear was measured. For 28-day-old specimens, depth of wear for the crazed surface was comparable to that of another steel-troweled surface without craze cracks. The researcher noted that when a 90-day-old craze-cracked specimen made with the same concrete was tested, it lost the shake-coat layer after only one test cycle of the abrasive wheel. This failure was attributed to a loss of bond between the high-shrinkage shake and the lower-shrinkage base concrete.

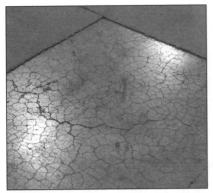

Photo A

Photo B

Photo C

Does this mean that slab surfaces with craze cracks are more susceptible to delamination or other failure modes? Based on observations of existing slabs, that seems unlikely. Photo A shows a craze-cracked convention-center floor that has received heavy hard-wheel traffic for many years. The floor is still in service and the craze cracks haven't affected its performance.

Photos B and C show the same exterior concrete slab in a northern climate. Photo B was taken in 1984 and Photo C in 1996. The slab has been exposed to freezing and thawing and deicing agents during the interim. Note that no further deterioration has occurred.

This anecdotal evidence supports the statement in Reference 2 that craze cracks don't affect the structural integrity of concrete and in themselves should not lead to subsequent deterioration of the concrete. There are many other examples of craze-cracked slabs that still perform their function, despite a possibly displeasing appearance.

Reference 2 further states: "Crazing is often self-healing and should not lead to problems of durability. If, however, the crazing is so severe that it leads for example to frost damage on a concrete paving, then it is likely that (a) it is so deep that it cannot strictly be called crazing and (b) the water content and the permeability of the concrete are so high that frost damage is inevitable even without the presence of cracks that look like crazing." To that latter statement, we say, "Amen."

References

1. Blake Fentress, "Slab Construction Practices Compared by Wear Tests," *ACI Journal*, July 1973, pp. 486-491.

2. *Non-Structural Cracks in Concrete*, Concrete Society Technical Report No. 22, The Concrete Society, 1982, pp. 26-27.

Reader response: Over the years, we have researched problems our customers have had with scaling of exterior concrete. Our winters are sometimes harsh, which causes an abundance of freeze-thaw cycles.

Recently, we have noticed some scaling in areas where the concrete surface is crazed. The mixes used for these areas contained at least 5½ sacks of portland cement per cubic yard and 5% to 7% entrained air. They were placed at slumps of 4 to 5 inches. In most of the scaled areas, the matrix covering the aggregate is ¹⁄₁₆ to ⅛ inch thick.

Everything we've researched on the subject of craze cracking suggests the cracks will not affect the structural integrity of concrete, But if you have cracks in exterior concrete, can't moisture penetrate into them, freeze and possibly scale the surface? Won't the hydraulic pressure created by the frozen moisture promote problems such as scaling? If the answer to these questions is yes, we suggest that crazing can lead to surface defects, which can indeed affect the structural integrity of concrete.

Clay Allen, Central Pre-Mix Concrete Co., Spokane, Wash.

Slab deflection: How much is too much?

Q. *As part of a large warehouse restoration project, we're repairing several deteriorated floor joints. The joints spalled under forklift traffic because of slab deflection caused by curling. The new warehouse tenants will be altering the forklift traffic routes, and we believe that deterioration at joints along the new routes is inevitable because of similar slab deflection. We'd like to underseal some of these joints before they deteriorate to the point where more extensive repairs are required. Do you have any suggestions on how to determine which joints are good candidates for undersealing and which joints should be left alone?*

A. Of course, there isn't a straightforward formula for determining which joints will deteriorate. Experienced contractors will take into account the vehicle weights, tire types, and amount of traffic when making a decision. You often can feel joint deflection simply by standing on the joint as a forklift passes over it. This is a crude method for detecting deflection, but it shows that a problem exists.

To help make repair decisions, one repair contractor we spoke to rolls a forklift over the joint and measures deflection with a modified Benkelman beam. The contractor then categorizes the potential for distress based on differential movement from one side of the joint to the other and found that:

- Movements of 0 to 0.005 inch indicate excellent joint performance and little curling.
- Movements of 0.005 to 0.010 inch indicate acceptable joint performance and tolerable curling.
- Movements of 0.010 to 0.017 inch are in the gray area. Repairs may improve floor performance but may not be cost-effective. Without repairs, the floor might deteriorate twice as fast but could still last 10 years without requiring major work.
- Movements of 0.017 to 0.030 inch are severe enough to cause deterioration that's three to four times as fast as normal. Problems such as joint deterioration and breaking of any vehicle guidance wires could occur within two years.
- Joint deterioration is unavoidable when joint deflections reach 0.05 to 0.10 inch.

Causes for corner cracking

Q. *In December 1996 we placed a floor slab for a manufacturing plant, and a year later we were called back by the owner because one section had cracked badly. Most of the floor is 8 inches thick, but the part that cracked was a 6-inch-thick aisle adjacent to an 8-inch-thick slab for the plant's production area. Specifications required a 5000-psi concrete containing 658 pounds of cement per cubic yard. We placed the concrete on a 6-inch-thick stone base with no vapor barrier and used 6x6 W2.9xW2.9 welded wire fabric, as required by the specification.*

This slab was the last pour, and we left the job after completing it. We heard that there was construction traffic on the slab within seven days after it was placed because the doorway leading to the 8-inch-thick slab was blocked. The sketch shows the cracking pattern that developed in the first year after the building was turned over to the owner.

What causes this kind of cracking pattern?

A. Much of the cracking appears to be related to corner curling at contraction-joint intersections. Curling is usually caused by unequal drying shrinkage: It's greater at the top of the slab than at the bottom. This causes the corners to curl upward, leaving them unsupported. Traffic loads then cause corner cracks like those that have occurred at nearly every contraction-joint intersection (see drawing). The multiple cracking in some of the panels may be caused by restrained drying shrinkage or thermal contraction.

Because the floor was placed in the winter, some of the curling may have been caused by the top of the slab cooling faster than the bottom. In an unheated building, the top of the slab would contract more than the bottom, creating corner curling similar to that caused by un-

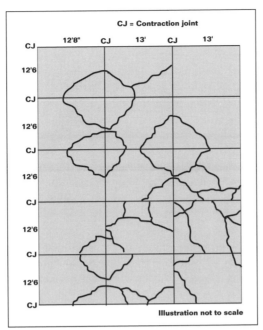

CJ = Contraction joint

Illustration not to scale

The cracking pattern in this 6-inch-thick slab indicates that the cracking probably is due to corner curling at contraction-joint intersections.

even drying shrinkage. If the slab was loaded after only seven days, the combination of unsupported corners and low strength at this age would increase the likelihood of corner cracking.

Q. *At a building we built 4 years ago, water appears on the sidewalk near one entrance every morning during the cooler fall months. It usually freezes and is a slipping hazard. The condition is localized and as soon as the temperatures get colder into the winter, the problem disappears. Another entrance 250 feet away has no condensation or icing problem.*

Before I learned about the problem, the building manager put a sealer on the concrete because he thought water was coming up through it. I covered an area with a plastic sheet, though, and found that moisture was condensing on the surface. The sealer keeps it from soaking into the concrete so the problem is worse than ever.

Is there any way to prevent this condensation? The owner is concerned about possible slip-and-fall injuries to people using the building but doesn't want to use deicing salts on the concrete.

A. Either of two relatively low-cost approaches might work. Diamond grooving would provide channels for the water and perhaps allow it to run off instead of freezing on the surface. Or mechanically scarifying the surface might also give enough texture to skidproof it even if some ice forms in the valleys. If you try scarifying, it also might help to stain the concrete a darker color so it absorbs more heat from sunlight.

Q. *We put in the foundation and floor slab for a house about 3 years ago. The floor slab performed fine for 2 years but then started heaving and cracking. Some of the cracks are up to ⅛ inch wide. Could water pressure have caused the heaving and cracking? The basement looks dry but last year we got lots of rain.*

Condensation causes slippery sidewalk

Delayed cracking of a basement floor slab

A. If the basement is dry it's unlikely that water pressure caused the cracking. If there was enough head to lift the concrete slab you'd probably see water at the joints or cracks. Swelling soil is a more likely cause. If the house was built over swelling soil during a dry period, a later wet spell could cause enough swelling to heave the slab and cause the cracking. Have a local consulting engineer look at the slab. He may know whether or not there are expansive soils in the area and can take samples to determine if they're present beneath the slab.

Plastic shrinkage crack prediction

Q. *I'm familiar with what causes plastic shrinkage cracks—the evaporation of moisture from the surface. I also know some ways to prevent their occurrence. What I haven't been able to figure out is when they will occur. Can you give me some guidelines so that I can be prepared with the manpower and materials needed?*

A. Plastic shrinkage cracking is more likely to occur on days when one or more of the following weather conditions are present: high temperature, low humidity, or wind. These conditions all lead to a high rate of evaporation of moisture from the surface. Use the following chart to estimate the rate of evaporation. Precautions against plastic shrinkage cracking should be taken when the evaporation approaches:

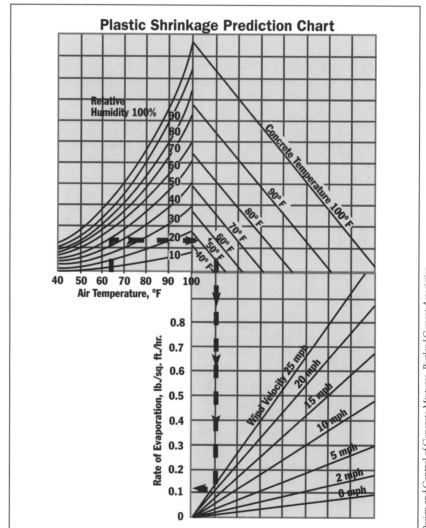

Plastic Shrinkage Prediction Chart

Relative Humidity 100%

Concrete Temperature 100° F

Air Temperature, °F

Rate of Evaporation, lb./sq. ft./hr.

Wind Velocity 25 mph

Design and Control of Concrete Mixtures, Portland Cement Association

- 0.20 lb/ft²/hr for portland cement concrete
- 0.15 lb/ft²/hr for portland cement concrete containing more than 15% pozzolan
- 0.10 lb/ft²/hr for portland cement concrete containing more than 5% silica fume

To use this chart:

1. Enter with air temperature and move up to relative humidity.
2. Move right to concrete temperature.
3. Move down to wind velocity.
4. Move left; read approximate rate of evaporation.

Chapter 8

Defects–Vertical Surfaces

Q. *We've got a dispute on a poured wall placement regarding acceptable bug hole size. ACI 116R-90, "Cement and Concrete Terminology," defines bug holes as small regular or irregular cavities usually not exceeding 15 mm (0.60 inch) in diameter. Does that mean any bug hole larger than 15 mm is unacceptable?*

A. ACI 116R-90 isn't a specification or standard. It simply describes bug holes and is silent on the subject of acceptable size.

In some specifications, permissible bug hole size is related to the end use of the concrete. Current U.S. Army Corps of Engineers guide specifications require patching ½-inch-diameter or larger bug holes in Class A surfaces (those prominently exposed to public view, where appearance is of special importance). But for Class C surfaces (permanently exposed surfaces not requiring other finishes), only 1½-inch-diameter bug holes or larger must be patched.

Naturally, contractors want to avoid patching bug holes because of the cost. Owners may also want to avoid patching for aesthetic reasons. *ACI Manual of Concrete Inspection* contains the following advice: "Consider whether the repair will be less apparent and have a more pleasing appearance than the original blemish. Unfortunately, many repairs do not. For example, surface voids (bug holes) are common and their repair, except for sack rubbing or similar treatment, may be less satisfactory than no repair."

Unfortunately, many specifications don't clearly state an acceptable size for surface cavities, and disputes can arise when owner's or architect's expectations about formed surface appearance are unrealistic. Help is on the way, however. The American Society for Concrete Construction is preparing a *Guide for Surface Finish of Formed Concrete* to help contractors and owners agree, in advance, on what constitutes an unacceptably large surface void on formed surfaces. The report should be available in 1998.

Bug holes: how big is too big?

Q. *We have a problem with too many bug holes on wall surfaces formed with rubber-lined wall forms. Are there ways to achieve the same smooth surface we can get with bare medium-density-overlaid plywood wall forms?*

A. We're not sure we know what you mean by "rubber-lined" wall forms. Are the forms lined with rubber sheets or mats as described in *Formwork for Concrete*, published by the American Concrete Institute? Or are you referring to an elastomeric form liner?

In any case, rubber or elastomeric liners are more impervious than MDO plywood and could produce more bug holes if all other conditions were equal. Because it's more pervious, an MDO form absorbs water and permits the passage of air and water from the concrete, thus reducing the size and number of bug holes.

Bug holes on rubber-lined wall forms

Form-surface absorbency also affects the thickness of the layer of form-release agent. A rubber or elastomeric surface, which is impermeable, won't absorb any form-release agent; an MDO surface, which is semi-pervious, will absorb some. So if you apply the same amount of form-release agent to both surfaces, a thicker coating will result on the rubber surface. Thick layers of release agent increase bug hole size and number. Ask the rubber-liner manufacturers to recommend a release-agent type and application rate.

It's also possible that the rubber surface is rougher than the MDO surface, making it easier for bubbles to cling to the surface. And if the rubber liner is installed to impart texture or pattern to the concrete, as is often the case, the surface area of the liner may be two to three times that of a flat form, requiring significantly more vibration to achieve a satisfactory surface.

Bug hole repellent

Q. *In our plant we manufacture manholes, inlet boxes, utility vaults, median barriers, and other concrete products. Most of our products (but not the median barriers) have small surface voids. We have always had such voids and we don't think that they cause any problems, but we seem to have more voids now than we had in the past. Could the superplasticizer we're using cause the increased number of surface voids? If so, what can we do to minimize the number of such voids?*

A. Superplasticizers may or may not cause more surface voids (known as bug holes) in wet-cast products. Superplasticizers tend to make concrete mixes "stickier," and that may contribute to increased voids. However, because concrete with a superplasticizer seems to flow so easily, there is a tendency to undervibrate the concrete. In the photo, there appear to be more large bug holes in a vertical line near the center. That seems to indicate that internal vibrators were used and the spacing of the insertions was too far apart. (Note: see the May 1996 *Concrete Journal*, page 318, for proper vibrating methods.) Other causes of bug holes might be using an excessive amount of form-release agent or an inadequate mix design—one that's too sticky.

Excessive bugholes may be caused by several production techniques, but probably not by a superplasticizer alone.

Q. *On some recent small concrete wall jobs, we found small surface air voids, or bug holes, on the surfaces of the hardened concrete. Although the walls seem structurally sound, I'm concerned about their appearance. Is there anything I can do to prevent bug holes from forming?*

A. Bug holes on the surface of concrete walls are usually caused by lack of consolidation at the concrete surface, which can leave air voids trapped next to the form. Sticky or stiff concrete mixes that tend to have low workability and may have an excessive sand or entrapped air content can be harder to consolidate.

To minimize surface air voids, insert the vibrator as close to the form as possible, but without touching the form. Proper vibrator spacing and duration of vibration also result in better consolidation. Avoid using vibrators with too large an amplitude, and completely insert the vibrator head. Also avoid applying high-viscosity form coatings in thick layers that can trap air and water.

Q. *Due to a wall form blowout on a poured wall foundation, one section of the wall about 4 feet long has a 2½-inch bow. Is there a faster way of removing the bow than grinding it down?*

A. Try sawing vertical slots about ½ to 1 inch apart in the bow. Vary the slot depth, making it deepest in the fat part of the bow and tapering down to shallower cuts near the edges. Then use a chisel or a small chipping hammer to remove the concrete. Making some horizontal cuts may speed the process. Then patch the surface with a concrete patching mortar.

Q. *On a hot day in August, we poured 40 lineal feet of concrete foundation wall for a house addition. The next day when we stripped the 2-foot-wide plywood-faced form panels we noticed about 10 to 15 cracks at the top of the wall. As shown in the photos, the cracks went from face to face but didn't penetrate to the full wall depth. They terminated at a depth of about 3 to 4 inches. What causes these cracks?*

Figure 1. Crack coinciding with the edge of a form panel.

Figure 2. Crack near an anchor bolt.

A. In one of your photos (Figure 1), the crack coincides with the edge of a 2-foot-wide form panel. Cracks at these locations may be caused by plastic settlement of the concrete over form ties.

However, your other photo (Figure 2) shows a crack near an anchor bolt but not at the edge of a form panel. If the temperature dropped over night, it's possible that this crack was caused by thermal contraction. Plastic shrinkage cracking is another possibility.

Are wall streaks caused by bad concrete?

Q. *A subcontractor built 350 lineal feet of 10-inch-thick foundation wall that is 8 feet high. When the forms were stripped, there were streaks that looked like veins on the surface of the concrete. The subcontractor believes they were caused by the aluminum forms, which were being used for the first time. The owner thinks the concrete is bad. Were the veins caused by the forms or by bad concrete?*

A. Veining has been reported to occur when calcium hydroxide released during cement hydration reacts with a new aluminum surface. The reaction produces hydrogen gas that rises to the top of the forms, producing the vein-like pattern. This usually happens only with new forms before an oxide surface builds up on the form surface.

Because the reaction occurs only at the surface, it doesn't affect the strength of the wall, but it can be unsightly. Two ways have been suggested to control the problem:

- Before using new aluminum forms, tell the contractor to brush on a lime (calcium hydroxide) and water paste, allow time for gas to form, and then wash off the paste. Repeat the treatment until no more gas bubbles form.
- As an alternate, have the contractor use an 80-weight gear lube as a release agent when the forms are first used. The gear lube has a consistency between oil and grease. It can be rolled on the new forms. After the first or second use, a normal form release agent can be used.

Chapter 9

Durability

Q. *What would cause a well-applied, good-quality clear sealer to peel off a concrete slab in thin sheets? Our crew applied the sealer after thoroughly cleaning the slab, following all of the manufacturer's recommendations. The slab is part of an interior loading dock with a hard-troweled finish. Concrete was placed directly on a vapor barrier with no granular layer on top of the vapor barrier. There is a sand fill beneath the vapor barrier. The slab was about a month old when the sealer was applied. This is all the more puzzling because we applied the same sealer, in the same way, to a similar slab about 30 miles away and had no problems with it.*

Both slabs were placed by the same subcontractor, with the same mix specified; however, the concrete came from two different suppliers. The same crew (not the slab subcontractor's) applied the sealer to both slabs at the same concrete age—about one month—and neither slab had any curing compound applied. The site for the problem slab was poorly drained, and there were numerous delays during construction due to moisture conditions.

The second slab, where the sealer is successful, is outdoors and has a rougher finish. This site was well-drained, and the slab was placed a little later in the season when rain was not as much of a problem.

A. We've heard of a similar problem occurring when nonbreathing polymer coatings are applied to concrete slabs. Blisters appear, caused by movement of water vapor to the concrete-slab surface (see *Concrete Construction*, June 1996, pp. 480-486). However, we've never heard of this happening with a sealer.

Reader response: The contractor does not state whether the sealer he applied was water- or solvent-based, but I assume that it was solvent-based. Although conditions such as surface preparation, air temperature, age of the concrete, and surface finish are critical to the successful application of a concrete sealer, a condition often overlooked is the dew point at the time of application.

The *dew point* is the temperature at which condensation of water vapor occurs. On a typical humid summer day in the Midwest, it is not uncommon for the dew point to be only a few degrees lower than the air temperature. Concrete substrates enclosed by new buildings are usually shaded and receive little air movement. As a result, substrate temperatures often are near or below the dew point.

Usually, the first coat of sealer will go down with no delamination problems, since it can adhere to a relatively rough or porous substrate, even if the substrate is moderately moist. As the first coat dries, solvent from the sealer leaves the film and cools the concrete substrate a few degrees, probably just enough to lower the substrate temperature below the dew point. You may have experienced this cooling effect if you accidentally spilled on your skin acetone or any solvent with a high speed of evaporation.

After the solvent has evaporated from the sealer, water vapor in the air will begin to condense on the surface of the first coat. Also, the first coat of sealer has penetrated the surface pores of the concrete, dramatically reducing the substrate's porosity. This decrease in porosity and the presence of condensed water on the surface interfere with the adhesion of subsequent coats of sealer. As we all know, water and oil (or solvent-based sealers) don't mix. The second coat of sealer will form on top of the first coat, with the condensed water sandwiched, or trapped, between the two. Since the second coat did not adhere to the first coat, it will delaminate in thin sheets, as the contractor described.

Cyler F. Hayes, Dayton Superior Corp., Chemical Operations, Oregon, Ill.

I have found this to be a common problem, usually caused by troweling the slab surface too hard, thereby bringing excessive laitance to the surface. The sealer adheres only to the laitance and soon flakes off.

I have also encountered this problem in a slab containing excess moisture. The slab was subjected to freezing conditions, and moisture that rose to the top of the slab condensed and froze, lifting off the sealer. However, the problem wasn't noticed until long after the sealer was applied.

Michael E. Shanok, Forensic Engineering, PC, Hamden, Conn.

Best material for embedded sleeves

Q. *We have to embed metal sleeves in a concrete floor slab and are concerned about corrosion. The sleeves protrude 4 inches above the floor surface. What's the best metal to use for this purpose?*

A. If no chloride-containing admixtures are used in the concrete and the floor isn't in a corrosive environment, steel sleeves should work fine. If the floor is in a corrosive environment, use either stainless steel or heavy-duty plastic sleeves.

Durable concrete for ice skating rink

Q. *We're building an outdoor basketball court that will be flooded in the winter and used as an ice skating rink. What can we do to protect the concrete from damage caused by freezing?*

A. This exposure isn't as severe as subjecting the concrete to deicing chemicals, which increase the number of freeze-thaw cycles. However, you should take the same approach that is used to produce a durable concrete driveway. Order a 4000-psi air-entrained concrete with a 3- to 5-inch slump, and specify an air content of 5% to 8% for ¾- or 1-inch maximum-size aggregate. Also specify chert-free coarse aggregate if it's available.

Don't start finishing until bleeding has stopped and the water sheen on the surface has disappeared. If a troweled surface is required, it's especially important to delay troweling until bleeding has stopped. This avoids trapping bleedwater beneath a surface sealed by troweling. Moist cure the concrete at least three days, then allow the surface to dry for at least 30 days before it's exposed to freezing.

Best coating for embedded aluminum conduit

Q. *"Building Code Requirements for Structural Concrete" (ACI 318-95) prohibits embedding aluminum conduit in structural concrete unless it's effectively coated or covered to prevent aluminum-concrete reaction or electrolytic reaction between aluminum and steel. What coatings are effective and where can I obtain coated conduit?*

A. A laboratory study conducted by the Portland Cement Association (PCA) in 1965 showed that some coatings made of lacquer or bitumen effectively protected aluminum conduit embedded in concrete and electrically coupled with embedded steel. However, the researchers cautioned that the coating performance under field conditions hadn't been evaluated. We couldn't find any recent studies of aluminum conduit coatings.

We did find a company that coats aluminum conduit with a 45-mil thickness of polyvinyl chloride. You can obtain a product catalog from Perma-Cote Industries, 29 Industrial Dr., Lemont Furnace, PA 15456 (412-628-9700).

Reference

G.E. Monfore and B. Ost, "Corrosion of Aluminum Conduit in Concrete," *Journal of the PCA Research and Development Laboratories*, January 1965, pp. 10-22.

Q. *We did concrete curb and gutter work for a school parking lot. Concrete was supposed to be air-entrained but it was a long haul to the job and the concrete lost air. We placed it anyway. Now the owner says we have to rip out $5,000 worth of curb and replace it. Are there any alternatives to this? Can't we seal the concrete to protect it from freeze-thaw damage?*

A. Most concrete people believe that sealers won't protect inadequately air-entrained concrete from freeze-thaw damage. However, some manufacturers' tests on non-air-entrained concretes protected by sealers show good resistance to scaling after 285 cycles of freezing and thawing (see *Concrete Construction*, October 1989, page 900).

Why not tell the owner you'll apply a sealer and will guarantee the work against scaling for 5 years? The sealer might prevent damage and will certainly cost a lot less than removing and replacing all the concrete. If the concrete does scale within a few years, you may still be able to repair damaged sections at a lower cost than for removal and replacement.

Q. *I understand the effects chlorides have on reinforcing steel in concrete. What effects, if any, do chlorides have on plain, nonreinforced concrete, which is typically used in sidewalks and driveways?*

A. Properly designed concrete mixes and careful attention to placing and finishing should produce a plain concrete slab that is resistant to chlorides. In some exposed concrete slabs, the chlorides in deicing chemicals have attacked the surface of the concrete and led to surface scaling. In most cases, this situation has been caused by low air content, overfinishing, or finishing with rainwater or bleedwater on the surface. A plain concrete slab with adequate entrained air and a low water-cement ratio at the surface should resist scaling from deicing chemicals. Eliminating birdbaths and low spots in the surface of the slab during placement also contributes to good performance. Allowing concrete to cure thoroughly also can improve resistance to scaling. After the concrete has dried for at least 28 days, apply a sealer to help keep out salt and moisture during service.

Q. *We're replacing concrete floors, pads, and pedestals damaged by hydrofluoric acid spills in an oil fuel processing plant. The acid is at ambient temperature and concentration varies from 10% to 60%. How do we protect the new concrete so it doesn't deteriorate?*

A. Bob Schoenberner of Schoen Industries, Inc., suggests that you use a vinyl ester material. Hydrofluoric acid attacks silica, so don't use a silica filler. The least expensive approach is a coating of carbon-filled vinyl ester. However, if the concrete cracks, the coating will probably

Saving a non-air-entrained concrete curb

Effect of salt on nonreinforced concrete

Protecting concrete exposed to hydrofluoric acid

crack too and allow acid to penetrate. To prevent this problem, first put down a glass-fiber-reinforced asphalt membrane and then apply the coating over the membrane. Instead of using a coating, you can trowel on a ¼-inch-thick carbon-filled vinyl ester mortar over the asphalt membrane. This provides more protection but costs more than a coating. For severe exposure conditions you may need to cover the membrane with carbon bricks set in a vinyl ester mortar.

Cause of surface deterioration

Q. *We placed a 6-inch-thick, 16x40-foot air-entrained concrete slab during cold weather. The 12-cubic-yard load of 3000-psi concrete was made with hot water and contained an accelerator that permits placement at temperatures down to 20° F. Concrete arrived at 9:30 a.m. with a 4-inch slump, and we added no additional water at the jobsite. We finished placing it an hour and 40 minutes after the truck had left the batch plant. High air temperature during the day was 42° F and low was 24° F. Although we waited until 7 p.m., the concrete still hadn't set, so we covered it with plastic sheeting, 1 foot of straw, and another piece of plastic sheeting. We were finally able to power trowel the slab at 2 p.m. the next day. At 7 p.m. that day we covered the slab again with plastic sheeting and 18 to 24 inches of straw. When spreading the straw, workers noted that the surface still wasn't real hard.*

On the second day following the pour, the weather warmed up and there was lots of fog, so we uncovered the slab. Snow fell soon afterward, and we couldn't work for 25 days. When we came back to the jobsite we removed 2 to 3 feet of snow and found that much of the slab was covered with 3 to 4 inches of ice. We used potassium chloride to melt the ice so it could be scraped off. Workers noticed that the concrete surface looked milky and seemed to be covered with a cream-colored cement film. They used squeegees to remove the rest of the deicer, but there was no water, so they couldn't wash off the surface. The weather turned cold again three days later and 14 more inches of snow fell.

When we came back to the job two weeks later and shoveled snow off the slab, most of the slab surface was badly damaged (see photos). In some places you could scrape off concrete to a depth of ¾ inch. Was this deterioration caused by the potassium-chloride deicer?

A. The use of a chloride deicing agent usually doesn't cause severe deterioration in such a short time. However, conditions were quite harsh because the concrete had no time to dry before being exposed to freezing and deicer application. Even air-entrained concrete may not survive the extreme conditions you describe. It's also possible that the fresh concrete froze.

To pinpoint the cause, you need to have cores taken and then examined by a petrographer for evidence that the fresh concrete froze. Chloride content, air content, and quality of the hardened concrete could also be estimated as an aid to finding out why the concrete deteriorated so rapidly.

Q. *We're applying a two-component polyurethane coating to the walls and floor of a new concrete reservoir. It's a high-build, 100% solids coating sprayed 125 to 140 mils thick in one pass. We first sandblast the concrete surface and repair any bugholes with a polymer-modified cementitious mortar. Then when the concrete is at least 30 days old we prime with a single-component aromatic urethane coating. We spray on the two-component urethane at a temperature between 140° and 160° F. Our problem is pinholes, or what our workers call "glass eyes." These are voids right beneath the surface, covered by a thin translucent membrane of polyurethane that breaks easily. Is this caused by gases coming out of the concrete? How can we correct the problem?*

Pinholes in polyurethane coating

A. We asked Jim Kubanick of Coatings for Industry Inc., to answer your question. He says that all urethane coatings cure by reacting with moisture in the air. The byproduct of this reaction is carbon dioxide. When the urethane is applied at high temperatures, or at moderate temperatures and high relative humidity, bubbling may occur. Adding urethane-grade solvent to the polyurethane (up to 20% by volume) will help to prevent bubbling but it will make the urethane sag more when sprayed on vertical surfaces. You may have to apply the thinned material in two layers instead of one. Also, make sure that both the concrete and the bughole repair material have fully cured before you apply the urethane.

Q. *Can you explain the difference between a sealer and a sealant?*

Sealers and sealants

A. There is confusion about sealers and sealants due to inconsistent descriptions of these materials in articles and in some manufacturers' literature.

Sealers are products applied to surfaces, usually to reduce the penetration of undesirable materials, such as water or chlorides. Sealers may penetrate into the surface or form a very thin surface layer (typically less than 3 mils). Thicker surface treatments usually are called coatings.

Sealants, sometimes called caulks, are used for filling voids, gaps, cracks, and joints. A variety of sealants are available for different applications and types of construction. Sealants may be rigid or flexible, and nonsagging or self-leveling, depending on performance requirements. A related term, seal, describes a gasket, strip, or gland, usually between panels or at isolation or expansion joints. Most seals are prefabricated, but some are field-molded.

Q. *We just finished an industrial floor job that meets flatness and levelness requirements and has no random shrinkage cracks. However, there is some craze cracking. The owner wants proof that these cracks won't get progressively worse with time and affect floor performance. Have any studies been done to show the effect of craze cracking on floor abrasion resistance or long-term performance?*

Effect of crazing on long-term floor durability

A. We don't know of any studies that show the effect of craze cracking on floor durability. We do know these cracks don't penetrate very far below the surface and they're so narrow that they're almost invisible. You might find an older industrial floor that's still performing well even with craze cracks present. Then show it to the owner of your floor to prove that floors with craze cracks can still perform well. Craze cracks are very common on many floors that receive hard trowel finishes so you shouldn't have much trouble finding an older floor that has the cracks.

Waterproofing concrete storm shelters

Q. *We build concrete shelters for farmers who want protection from tornados. The shelters have a 10x10-foot footprint and walls are 7 feet high. We cast the floor slab first, leaving a keyway at the edges, then cast the walls and roof in one pour. The shelter sits in the ground with about 1 foot of the wall exposed and there's a metal door in the roof.*

Here's the problem. We're getting moisture leakage into the shelters, making them damp and uncomfortable. Sometimes it leaks in around tie holes and sometimes at the joint between the floor and walls. In one case we put a 6-mil vapor barrier on the ground and brought it up around the edges of the slab but still got water coming up through the floor.

I have two questions. How can we economically waterproof the shelters we've already built so the owners will be satisfied? We don't want to dig if it can be avoided. And how can we build shelters that don't leak? If we put in granular material and drain tiles at the bases of walls, what do we drain the water to?

A. If the shelters periodically dry out, wait for a dry period and use a so-called negative side waterproofing product that can be applied from the inside. Polymer or cementitious products can be applied by a do-it-yourselfer. Follow manufacturers' instructions. A list of waterproofing material manufacturers can be found in the *Concrete Construction* buyers' guide company directory. If the shelter is perpetually wet, you might solve the problem by injecting a hydrophilic foam through holes drilled through the concrete from the inside near the source of the leakage. You'll probably have to hire a specialty contractor to tackle this job and cost will be higher than for using surface-applied products.

To prevent leakage during new construction you could try several approaches. Instead of using drain tiles and granular material, you could place the concrete slab on bentonite panels, use bentonite rope in the keyway between floor slab and wall, and apply bentonite panels to the walls before backfilling. You could also use adhered or loose-laid sheet membranes or liquid-applied solvent systems. With these methods you're trying to keep water out even if it rises and puts a liquid head on the floor and walls. Obviously, the floor and walls have to be stout enough to resist the pressure caused by a liquid head.

If you use underdrains and granular material or drainage boards to direct water away from the shelter, you'll need to dig a separate pit below the level of the shelter floor, fill the pit with granular material, and run the drainpipe to the pit.

Effect of petroleum products on concrete

Q. *Does exposure to petroleum-based lubricating oils and transmission fluids cause concrete floors to disintegrate?*

A. Pure mineral oils such as gasoline, fuel oils, lubricating oils, and petroleum distillates reportedly don't attack mature concrete (Refs. 1 and 2). However, a book published in 1950 (Ref. 3) says that lubricating oils are often improved by adding fatty oils (animal and vegetable oils). These oils can decompose to form fatty acid, which disintegrates concrete. We don't know whether modern lubricating oils contain fatty oils, but they do contain additives that improve the performance of refined-petroleum base stocks. We don't have any information on the effects of such additives on concrete.

It's also known that used lubricating oils have higher levels of acidity because of oxidation. However, the references cited don't address the effects of used oils vs. virgin oils. Because of their increased acidity, it's possible that used oils might attack mature concrete. But if that's the case, you'd expect to find a lot of deteriorating, oil-stained garage floors. Have any of our readers seen evidence of concrete deterioration in industrial floors repeatedly exposed to oil spills?

References

1. F.M. Lea, *The Chemistry of Cement and Concrete*, 3rd ed., Chemical Publishing Co. Inc., New York, 1971, p. 660.

2. Sandor Popovics, "Chemical Resistance of Portland Cement Mortar and Concrete," *Corrosion and Chemical Resistant Masonry Materials Handbook*, Noyes Publications, Park Ridge, N.J., 1986, p. 336.

3. A. Kleinlogel, *Influences on Concrete*, Frederick Ungar Publishing Co., New York, 1950, pp. 91-93.

Reader response: In the March 1997 *Concrete Construction* Problem Clinic (pp. 314-315), a reader wanted to know if exposure to petroleum products causes concrete floors to deteriorate. I have some experience that may provide a partial answer.

I am a specialist concrete building engineer and was retained to investigate severe concrete deterioration of reinforced-concrete foundations, piers and beams under a paper machine at a plant in British Columbia. The concrete had been placed in the late 1920s and may have used aggregates from a saltwater source. Throughout the years, lubricating oils from the machinery spilled over certain parts of the concrete, resulting in the deterioration, which had progressed several inches into the concrete. The concrete away from the oil exposure showed no signs of breakdown and was of good quality, with core strengths of about 3000 psi.

Thus, in this case, we know lubricating oils were the main contributor to the concrete deterioration. Conjecture is that the lubricating oils of bygone years (and perhaps the present) contain sufficient sulfur, which over time converted to acid in the alkaline concrete.

At another plant I investigated, where diesel injectors were being remanufactured, the concrete floor had extensively deteriorated in areas where diesel oil had been spilling continuously for about 20 years. The deterioration was due to chemical attack by the oil.

G.W. Spratt, Gordon Spratt & Associates Ltd., Vancouver, British Columbia

Producing acid-resistant concrete pipe

Q. *Our concrete pipe company has been asked to produce pipe to resist an acid solution with a pH of about 5.2. How can we calculate what the life expectancy of the pipe should be? Also, what admixture, if any, would make the concrete resistant enough to prolong its life? Would it help to use sacrificial concrete?*

A. Calculation is an unlikely way to get an answer because there are too many unknowns of temperature, flow rate, and the rate at which reaction products are removed and new surfaces exposed. We do not know of any admixture, other than a polymer used to make polymer concrete, that will significantly improve the acid resistance of portland cement concrete. Fly ash and other pozzolans will help, but will probably not give the amount of increased life needed. Switching from portland cement to calcium aluminate cement also might be helpful.

Sacrificial concrete (made by using limestone as an aggregate) might prolong the life to some extent, but possibly not to the extent needed in a flowing system where there is an essentially inexhaustible supply of acid.

Perhaps the best way to ensure pipe longevity is to use a coating or liner. ACI 515R-79, "A Guide to the Use of Waterproofing, Dampproofing, Protective, and Decorative Barrier Systems for Concrete," provides additional information.

Reader response: In your answer to the question on acid-resistant pipe (*Concrete Journal*, November 1994, page 40), you missed two important facts in the field of chemically resistant concrete. First, the Tennessee Valley Authority has recently completed scientific experiments of their own device to determine agricultural chemical corrosion on concrete. Second, sulphur concrete has been proven to be of high chemical resistance. Sulphur concrete, though, has failed to gain widespread acceptance because production requires specialized equipment and materials not common in the ready mix market. Sulphur concrete may actually be better-suited to the precast industry than to cast-in-place.

Robert D. Dienstbach Jr., WESTEC Barrier Technologies, St. Louis

Galvanized rebar

Q. *I understand that the products of reinforcing steel corrosion occupy a greater volume than the original steel. This volume increase exerts tensile stresses in the concrete, causing it to spall. We're considering using galvanized reinforcing steel in which a zinc coating acts as a sacrificial anode, protecting the steel. However, I've heard that zinc also expands when it corrodes. Wouldn't this cause the same problem as steel? Also, what would happen if, rather than using galvanized rebar, you simply added finely divided zinc powder to the mix? Would the rebar be protected?*

A. Depending on the oxidation state, the corrosion products of steel can occupy more than 6 times the volume of the original steel. Also, the corrosion products of steel are highly insoluble in concrete. On the other hand, the usual zinc corrosion product, zinc oxide, occupies only 1½ times the volume of zinc. In addition, the zinc corrosion products are more soluble in the concrete environment, so they may diffuse some distance from the metal-concrete interface. Therefore, zinc corrosion induces much lower tensile stresses in concrete than steel corrosion.

Adding zinc powder to the concrete mix would not provide cathodic protection to the rebar. There must be direct electrical contact between the zinc and steel to provide an avenue for electrons to flow from the anode (zinc) to the cathode (steel).

Reader response: We have found that corrosion of zinc in a high-chloride concrete can produce zinc hydroxy chloride, which we calculate as occupying almost four times the volume of zinc. We have found that development of this corrosion product can crack concrete.

William G. Hime, Principal, Erlin, Hime Assc. Div., Wiss, Janney, Elstner Assc. Inc.

Protecting concrete from hydrochloric acid

Q. *We're replacing a concrete pad for a storage tank that contains hydrochloric acid. The pad is deteriorated badly due to spillage and the new pad will also be exposed to spills. Is there a special concrete we can use or a protective sealer?*

A. A sealer alone won't do the job. You'll probably need to use a protective membrane. Bituminous materials or sand-filled epoxies, polyesters, or polyurethanes can be used in a barrier system.

It's also a good idea to use low permeability concrete so that even if the membrane fails, deterioration will be slowed. The concrete should have a low water-cement ratio (0.40 or lower). Use silica fume as an admixture to lower permeability even more. It may help to use a siliceous aggregate (gravel) rather than a calcareous one (crushed limestone).

If the pad will be exposed to abrasive wear, you may also need to protect the membrane. Repair contractor Bob Schoenberner described a job where a ¼-inch-thick fiberglass-reinforced asphaltic membrane was installed first to protect the concrete from acid attack. Then the membrane was topped with a 1½-inch layer of fiber-reinforced, acid-resistant potassium silicate concrete. See *Concrete Construction* magazine, February 1989, page 95, for further details.

For more information about protective coatings, see ACI 515.1R-79 (85), "Guide to the Use of Waterproofing, Dampproofing, Protective, and Decorative Barrier Systems for Concrete." Order this 44-page report from the American Concrete Institute, P.O. Box 9894, Farmington Hills, MI 48333.

Sealers for driveways

Q. *What's the best sealer to use on driveways to reduce scaling or to make oil stains easier to remove?*

A. If the concrete is properly air-entrained, scaling should not be a problem. But if the concrete near the surface does not contain an adequate air-void system, scaling can occur. In these cases, ACI 332R-84, "Guide to Residential Cast-in-Place Concrete Construction" recommends applying two coats of boiled linseed oil that is thinned with an equal amount of mineral spirits, turpentine or naphtha. But Ed McGettigan, chairman of ACI Committee 515, "Protection Systems for Concrete," says that linseed oil darkens the concrete and needs to be reapplied every year or two.

To reduce scaling by preventing water from entering the concrete, McGettigan recommends applying either a silane or a siloxane sealer with an active ingredient content of 10% to 20%. These sealers should provide effective water repellency for three to five years, but they will not prevent oil from soaking into the concrete. To repel oil, apply a low-solids (about 15%) acrylic or urethane sealer, which should last from two to five years. Acrylic and urethane sealers will give the concrete a glossy appearance and can make smooth concrete slippery.

Concrete sealers based on aluminum stearate are available, but their low active-ingredient content makes them less effective than silanes and siloxanes or acrylics and urethanes.

Footings exposed to frost

Q. *We're doing a renovation job on a residential basement. As part of the work, we're removing backfill in one area and cutting a 3-foot-wide door in the concrete foundation wall to create a walkout basement. Then we're adding a timber retaining wall to hold back the backfill on each side of the door.*

Because we've removed the earth, soil beneath the part of the footing at the door is exposed to freezing. Should we protect the footing by putting insulation beneath the concrete sidewalk leading away from the door?

A. We asked Kansas City, Mo., consulting engineer Willard Norton to answer your question. He says that insulation isn't needed. Only a small length of the footing is exposed, and there will probably be enough heat loss from the basement at the doorway to prevent ground freezing.

Will dry-mix shotcrete resist freeze-thaw cycles?

Q. *We have a repair job for exterior concrete where we plan to use dry-mix shotcrete. The structure will be exposed to freezing and thawing. The engineer says air can't be entrained in dry-mix shotcrete and he questions whether the repairs will be durable under freeze-thaw conditions. Are there any research data that will ease his concerns?*

A. Several studies have shown that dry-mix shotcrete is indeed resistant to cycles of freezing and thawing. In one study conducted by the Portland Cement Association, shotcrete test panels were obtained from contractors known for quality workmanship. The shotcrete contained only cement, water, and sand. The 28-day compressive strength of cubes from all panels tested ranged from 7340 to 11,400 psi. Specimens sawed from the panels were frozen and thawed in fresh and saltwater.

Freeze-thaw durability after 300 cycles in fresh water was comparable to that of high-quality, air-entrained concrete. In saltwater tests, weight loss was high even though length change and relative dynamic moduli of elasticity were satisfactory. This led the researchers to suggest that for saltwater exposure a sealer or coating might be needed. They concluded, though, that dry-mix shotcrete can provide a high-strength durable shotcrete if made using sound materials, in the proper proportions, and applied by an experienced shotcrete crew.

In another study, straight cement dry shotcrete mixes and mixes containing added silica fume were subjected to freezing and thawing. Compressive strengths varied from 6080 to 7510 psi. After 300 cycles of freezing and thawing, test specimens were still in excellent condition. There was no significant scaling or surface deterioration and sawcut edges were still sharp.

Results of several other laboratory investigations are reviewed in Reference 3 below.

References

1. Seegebrecht, Litvin, and Gebler, "Durability of Dry-mix Shotcrete," *Concrete International*, October 1989, page 47. Available from the American Concrete Institute, P.O. Box 9894, Farmington Hills, MI 48333.

2. Morgan et. al., "Freeze-Thaw Durability of Wet-Mix and Dry-Mix Shotcretes with Silica Fume and Steel Fibres," *Cement, Concrete, and Aggregates*, Winter 1988, page 96. Available from ASTM, 1916 Race St., Philadelphia, PA 19103.

3. Morgan, "Freeze-thaw Durability of Shotcrete," *Concrete International*, August 1989, page 86. Available from the American Concrete Institute, P.O. Box 9894, Farmington Hills, MI 48333.

Does hot, windy weather affect silane performance?

Q. *Can weather conditions affect the amount of material that is absorbed when a silane water repellent material is sprayed on concrete? I've read that on hot, windy days, so much silane may evaporate that the treatment has little effect.*

A. Data published in an article in *Concrete International* (October, 1990) shows loss of material due to evaporation to be minimal. Author Edward McGettigan says absorption is a faster reaction than evaporation because capillary forces are stronger than any other participating factor. To prove this with test results, researchers heated concrete, brick, and stone specimens to a surface temperature of 176° F using quartz lamps. They treated the specimens at that temperature by pouring a silane solution directly on the surface. During the test period, samples also were exposed to a 7-mph wind. In absorption tests, water uptake of the treated specimens was then compared with that of untreated specimens. Untreated concrete specimens absorbed 15 times as much water as the silane-treated specimens. Similar results were obtained for the brick and stone.

Silanes vs. siloxanes

Q. *We've installed several patches in the traffic deck of a parking garage and are now looking to apply a clear penetrating sealer to the deck to protect it from water and deicing salts. I've received recommendations from several sources to apply either a silane or a siloxane. What's the difference between these two types of sealers?*

A. Silanes and siloxanes, the two most common penetrating sealers, are both derived from the silicone family. Despite being very closely related, they have significant performance differences.

Unlike silanes, which require a high pH to catalyze, siloxanes are not dependent on substrate pH. Because of this, siloxanes are ideal for treating brick, stucco, and stone.

Because silanes are made up of smaller molecules than siloxanes, they typically will obtain deeper penetration than siloxanes. As a result, silanes perform well under abrasion and weathering. A consequence, however, of this small molecular size is that silanes are relatively volatile. Therefore, the solids content of a silane product should be high enough to compensate for the loss of reactive material through evaporation during application and cure.

Siloxanes, because they are less volatile, generally offer good water repellent performance at lower initial cost than do silanes. However, for concrete surfaces subjected to abrasive wear such as pavements and decks, treatment with a silane sealer will provide longer lasting protection.

In regard to surface texture and color, treatment with silane sealers typically cannot be detected visually. Siloxane products may slightly darken the treated surface.

Reference
"Master Builders Construction Products Handbook," Master Builders Inc., 1993.

Q. *At a chemical processing plant, two areas of the concrete floor are deteriorating under the attack of chemical spills. One of the areas is exposed to an organic solvent; the other is exposed to an inorganic acid. What types of overlays can be applied to the floor to resist these chemicals?*

Which overlays resist acids and solvents?

A. Because of the widely varying formulations of many generic types of overlays, it's difficult to make generalizations about their chemical resistance. Typically, vinyl esters and furans are resistant to organic solvents, while epoxy novolacs and methyl methacrylates have moderate resistance. In addition, epoxies are resistant to some solvents.

Epoxy novolacs, furans, vinyl esters, and carbon brick all are resistant to inorganic acids. Potassium silicates and acid brick also are resistant, but not to hydrofluoric acid. In addition, sulfur cement and urethanes have moderate resistance to inorganic acids.

Reference
Peter H. Emmons, *Concrete Repair and Maintenance Illustrated*, R.S. Means Co. Inc., 1993.

Q. *Does an air-entraining agent serve any purpose in wet-mix shotcrete? I believe shotcrete is placed with such force that any air-entraining effect is negated. Should we use air-entraining agents in concrete that will be placed by wet-mix shotcreting?*

Is entrained air required in wet-mix shotcrete?

A. The ACI 506 "Guide to Shotcrete" recommends entraining air in wet-mix shotcrete. Section 2.7.2 of that document says: "Wet-mix shotcrete that will be exposed to freeze-thaw cycles should be air entrained. Air entrainment tends to make some mixes more workable and may reduce rebound. Data are available to show that a significant quantity of the entrained air in wet mixtures is lost in gunning, but this may be offset by increasing the amount of air-entraining agent."

Section 2.4 of ACI 506.2 "Specification for Materials, Proportioning, and Application of Shotcrete" also states that for wet-mix shotcrete exposed to freezing and thawing, an air-entraining admixture should be specified.

Will grooving weaken a floor?

Q. *We're considering grooving a slippery concrete floor in our manufacturing plant to reduce the danger of accidents. However, I'm concerned about the effect of such grooving on the floor's structural strength and abrasion resistance. Is this a valid concern?*

A. Larry Mosher, executive director of the International Grooving and Grinding Association, says this isn't a concern. Grooved bridge decks can withstand the forces applied by 18-wheelers, and many industrial floors have been grooved successfully.

In one case, chemical plant ramps with ¼x¼-inch grooves on 2-inch centers showed no signs of wear after several years of forklift traffic. Floors have performed well even when the grooves have been cut on 2-inch centers both ways, producing a square groove pattern.

Why 0.35 w/c for 4000-psi concrete?

Q. *When pricing concrete, our salespeople have a problem with projects for which there are conflicting specification requirements. Some engineers and architects in our area specify both compressive strength and maximum water-cement ratio, but the water-cement ratios are ridiculously low for the strength given. For instance, for a 4000-psi 28-day strength, the maximum water-cement ratio is 0.35 for air-entrained concrete and 0.44 for non-air-entrained concrete. What is the basis for these extremely low water-cement-ratio values? And is there any way that I can convince the engineer or architect that they don't have to be that low?*

A. The requirements appear in the August 1992 version of the American Institute of Architect's (AIA) *MASTERSPEC*. They're probably based on Table 3.10 of "Specifications for Structural Concrete for Buildings" (ACI 301-89) and the similar Table 5.4 in "Building Code Requirements for Reinforced Concrete" (ACI 318-89). However, the AIA values are a misuse of the tables cited.

The tables give values of maximum permissible water-cement ratios to be used when acceptable field test records or trial mixture data are not available. The tables do not apply when acceptable field test records or trial mixture data are available. Nor do the tables apply to concrete containing admixtures other than those used exclusively for the purpose of entraining air.

American Concrete Institute committees 318 and 301 recognized the problems caused by misuse of these tables and deleted them from the current versions of the respective documents ACI 318-95 and ACI 301-96. However, the requirements are still included in many specifications.

Unless these low water-cement-ratio values are needed to ensure adequate durability, you could tell the architect or engineer that such specification requirements can actually produce lower-quality concrete. That's because the increased cement contents contribute to higher drying shrinkage and heat generation. This, combined with a higher modulus of elasticity (stiffer concrete) increases the possibility of random cracking caused either by restrained drying shrinkage or thermal contraction.

Air-content specifications for exterior flatwork

Q. *Presently, our specifications for exterior flatwork call for a compressive strength of 3500 psi with 4% (±1%) entrained air. I've been told that some municipal agencies require up to 9% entrained air for exterior concrete exposed to freeze-thaw conditions. How much entrained air should we specify for sidewalks, patios, curb and gutter, and other exterior flatwork?*

A. The proper amount of entrained air to resist the effects of freeze-thaw is dependent on such factors as aggregate size and freeze-thaw exposure. Guidelines and recommendations for

Total Air Content for Frost-Resistant Concrete

Nominal Maximum Aggregate Size (inches)	Air Content (Percent) ACI 318		ACI 301
	Moderate Exposure	Severe Exposure	Destructive Exposure
⅜	6	7½	6-10
½	5½	7	5-9
¾	5	6	4-8
1	4½	6	3½-6½
1½	4½	5½	3-6
2	4	5	2½-5½
3	3½	4½	1½-4½

Note: The requirements of ACI 318 match the recommended total air content values of ACI 201.2R, ACI 211.1, ACI 345, and ASTM C 94. The required or recommended air contents depend on aggregate size and freeze-thaw exposure conditions (described below). ACI 318 specifies a tolerance of ±1½% on total air content. ACI 301 states a range and doesn't indicate a further tolerance.

Moderate exposure—Conditions in a climate where freezing is expected, but where concrete will not be continually exposed to moisture or free water for long periods before freezing, and will not be exposed to deicing agents or other aggressive chemicals. Examples include slabs that are not in contact with wet soil and will not receive direct applications of deicing agents.

Severe exposure—Conditions in which concrete is exposed to deicing chemicals or may become highly saturated by continued contact with moisture or free water before freezing. Examples include pavements, curbs, gutters, and sidewalks.

Destructive exposure—Conditions in which concrete is exposed to freezing and thawing, severe weathering, and deicing chemicals. (Similar to severe exposure conditions.)

the amount of air required can be found in ACI 301, "Specifications for Structural Concrete for Buildings," and ACI 318, "Building Code Requirements for Reinforced Concrete." These guidelines can be incorporated by reference when writing specifications.

The table summarizes and compares the suggested air contents. ACI 301 requirements apply only to concrete subject to destructive exposure (see definition in box) and are stated in a range without a tolerance. Some specifications allow a 1% reduction in air content when compressive strengths are greater than 5000 psi, because higher-strength concretes have greater frost resistance.

ACI 332,"Guide to Residential Cast-in-Place Concrete Construction," suggests air contents similar to ACI 318. ACI 332 includes a map of the United States to assist in determining exposure levels.

For more information about air contents for frost-resistant concrete, see the article "Specifying Air-entrained Concrete" (*Concrete Construction*, May 1993, pp. 361-367).

Q. *Three different strengths of concrete were specified for one of our jobs in the Midwest: 3000 psi for foundations, 4000 psi for flatwork and 5000 psi for structural members above grade. Specifications give the same maximum water-cement ratio for all three concretes—0.44. Working with our concrete supplier, we submitted proposed foundation mix proportions for the engineer's approval. Under pressure to maintain the schedule, we placed the foundations, though the engineer hadn't yet re-*

Is a 0.44 water-cement ratio needed for foundations?

sponded. About 30 days later he returned the submittal marked "not approved" because the water-cement ratio calculated out to 0.48. Are we in trouble, and if so, how bad? Why is the same water-cement ratio limit specified for three different kinds of concrete?

A. We don't know why the same water-cement ratio is specified for foundations, flatwork and structural members above grade. When both strength and water-cement ratio are specified, it's usually because the water-cement ratio needed for durability is higher than that needed for strength. But a water-cement ratio as low as 0.44 is seldom needed for foundation concrete. ACI 318-95, "Building Code Requirements for Structural Concrete," requires a maximum water-cementitious materials ratio of 0.45 for concrete exposed to freezing and thawing in a moist condition, deicing chemicals or high sulfate levels. It requires a 0.40 maximum for corrosion protection of reinforcement in concrete exposed to chlorides from deicing chemicals, salt, salt water, brackish water, seawater or spray from these sources. Since you're in the Midwest, foundation-concrete exposure to sulfates, salts or freezing and thawing is unlikely.

How much trouble are you in? If the owner tries to require you to remove and replace the foundation concrete, your cost would be high. However, you could argue that this extreme measure constitutes economic waste. In 1992, a U.S. Court of Appeals ruled that when a contractor substantially but not strictly complies with a contract's specifications, the owner may not require replacement of the work if replacement would amount to economic waste. The owner may only take a credit (see the article "Concrete Law: Ripping Out Work Ruled Economic Waste" in *Concrete Construction*, May 1994, p. 441).

You could argue that as long as strength is adequate, the owner hasn't been damaged by the slightly higher water-cement ratio. The owner would have to prove damage, and it would be very hard to produce data showing that foundation concrete with a 0.48 water-cement ratio is less durable than concrete with a 0.44 ratio, especially when exposure to sulfates, chlorides or freezing and thawing is unlikely. On that basis, you could ask for a use-as-is disposition of this specification nonconformance. Under the economic-waste rule, restoration of any damage to the owner would be limited to reimbursement of the cost of the cementitious material needed to produce a 0.44 water-cement ratio instead of the 0.48 ratio used. As with any legal question, consult with your lawyer before planning your strategy.

Reader response: The answer to the Problem Clinic question "Is a 0.44 Water-Cement Ratio Needed for Foundations?" (*Concrete Construction*, October 1997, pp. 836-837) is confusing and incomplete. It says that the American Concrete Institute requires a maximum water-cementitious materials ratio of 0.45 for concrete exposed to freezing and thawing in a moist condition, deicing chemicals or high sulfate levels. Then it says that since the job in question is in the Midwest, foundation-concrete exposure to sufates, salts or freezing and thawing is unlikely. Either you're not in the same Midwest I'm in, or you are referring to construction, not the environment.

Further, you may not be aware that a judge in California has ruled that he would hear of no concrete repair that did not correct the water-cement ratio. Tell me how you do that!

William G. Hime, principal, Erlin, Hime Associates, Northbrook, Ill.

Editor's response: Mr. Hime is correct that my response needs clarification. I was referring to the construction and the environment. The foundation in question is in the Midwest but not in the northern Great Plains area, where attack by sulfate-bearing groundwater has been reported. The caller made no mention of contract documents referring to requirements for sulfate resistance, so I considered sulfate attack unlikely. Because it's a foundation, I also considered it unlikely that the concrete would be exposed to attack by deicing salts or repeated freeze-thaw cycles while in a moist condition. Thus, I couldn't explain why a 0.44 water-cement ratio was needed.

As to the California judge's directive, it brings to mind a line from Oliver Twist: "If the law supposes that," said Mr. Bumble . . . "the law is a ass, a idiot."

Ward Malisch

Chapter 10

Finishing

Q. *Specifications for a job I'm doing call for a troweled surface for air-entrained concrete flatwork. I've heard that air-entrained concrete shouldn't be troweled. How can I convince the architect that troweling is a mistake?*

A. Air-entrained concrete can be troweled. If the concrete will be exposed to freezing and thawing, however, finishers must avoid overfinishing the surface. Overfinishing may drive air out of the surface layer and cause scaling of the concrete when it's exposed to freezing and de-icing agents.

You might ask the architect why a troweled surface was specified. Simply striking off the surface and brooming it reduces the chances of overfinishing and will produce the most durable surface for concrete exposed to freezing and thawing. And a broomed surface is less likely to cause slip-and-fall accidents when the surface is wet.

Q. *Recently, I saw a concrete patio that had a rock-salt finish, which is characterized by many small surface indentations. Can you explain how to produce such a finish?*

A. We recently watched Bob Harris from L. M. Scofield Co. produce such a finish at the World of Concrete '95 Mega Demos. Here's how it's done:

1. Place, strike off, and bull float the concrete.
2. Let the concrete bleed and the bleedwater evaporate.
3. Seed the concrete with coarse rock salt. This product can normally be found in hardware stores. It's used as a deicer during the winter months and is also used in water softeners. The more salt you put on the surface, the more voids you will get. It may take some practice to get the look you desire.
4. Cover the slab with 1-mil plastic sheeting.
5. Float and trowel on top of the sheeting. If the concrete is very stiff, you may need to tap the salt into the concrete.
6. After troweling, remove the sheeting.
7. After the concrete sets, saturate the surface with water to dissolve the rock salt. Often it helps to sweep the surface with a broom while it's saturated.

You may see dark rings around the pieces of salt. This is a discoloration due to the salt and should fade during washing and with exposure to sunlight.

This rock salt finish isn't recommended for exterior concrete flatwork in cold climates because it may be damaged by freeze-thaw cycles.

Air-entrained concrete can be troweled

Producing a rock-salt finish

Rock-salt concrete finish

Q. *A client of our New Jersey architectural firm wants to use a rock-salt texture for the concrete deck of her swimming pool. We're not familiar with this process and wondered how the effect is achieved.*

A. The texture is produced by scattering rock salt over the surface after floating. The finisher presses the salt grains into the surface with a float or roller, leaving only the tops exposed. After curing the slab for seven days under waterproof paper, workers wash and brush the surface, leaving pits or holes from the dislodged or dissolved salt grains. A slightly different procedure is described in the Problem Clinic article in *Concrete Construction*, May 1995, page 475.

A rock-salt finish isn't recommended for use in areas subject to freezing weather because water trapped in the holes tends to spall the surface when frozen. A more detailed account of the finishing process is provided in the Portland Cement Association publication *Color and Texture in Architectural Concrete*. You may order this book online or call the Aberdeen Bookstore at 800-323-3550.

Finish required for floor to receive a polyurethane coating

Q. *We're placing a concrete floor that will later receive a light-reflective polyurethane coating. Specs require a 1- or 2-mil rough finish. What does that mean? Do we just magnesium float it, give it a sweat trowel finish, or what?*

A. Give the floor a normal steel trowel finish. Urethane coatings have hardly any build. Even with three coats (primer, film build, and finish) the total thickness won't exceed 6 mils (0.006 inch). Thus the coating doesn't alter the finish you provide.

Can power trowel blades overlap hardened concrete?

Q. *I've read about false curl, which results when concrete settles in the middle of a strip pour, making the fresh concrete surface lower than the side forms. While floating and troweling the surface, finishers work material toward the edge, making it higher than the middle of the strip.*

To prevent false curl when power floating and troweling adjacent to a previously placed floor slab, why can't you allow a small portion of the blades to overlap the hardened concrete surface?

A. We can think of at least three reasons for avoiding this practice:
- If the fresh surface is too low, you could damage the float or trowel blades as they hit the edge of the hardened concrete.
- Even if the surface isn't low, paste will be smeared on the hardened surface when the blades pass from the fresh concrete to the hardened concrete. These smears are very hard to remove.
- The hardened surface is an elevation control for the newly placed slab. If the power trowel drags paste onto this surface, it also changes the elevation control, making the edge of the newly placed slab higher than intended.

Finishing steel-fiber-reinforced concrete

Q. *Are special techniques needed to finish steel-fiber-reinforced concrete to prevent fibers from protruding from the surface?*

A. Although most conventional equipment can be used to finish concrete with steel-fiber reinforcement, some different techniques are needed to get the best possible results, according to ACI 544, "Guide for Specifying, Proportioning, Mixing, Placing, and Finishing Steel Fiber

Reinforced Concrete." Flat, formed surfaces typically do not need special attention since they do not show fibers and are smooth after the forms are stripped. Chamfers and rounds at form edges and corners will keep fibers from protruding in these areas.

Slab surfaces usually require more attention. To provide added compaction and bury surface fibers, use a vibrating screed to strike off the concrete. A jitterbug can be used in areas where a screed is not practical, but be careful not to overwork the surface. Use magnesium floats to close the surface; wood floats can cause tearing. When using floats and trowels on steel-fiber-reinforced slabs, keep the tools flat because their edges can cause fibers to spring out of the surface. Do not use burlap drags for texturing, since burlap tends to lift the fibers and tear the concrete surface.

Q. *We're working with a specification that calls for final troweling until the trowel makes a ringing sound as it moves over the surface. This is to be followed by scarifying the surface with a soft-bristled broom. I don't think the broom will scarify a hard-troweled surface, but is there anything in writing to prove this?*

A. You're correct. We talked with two experienced concrete finishers, Carl Peterson, former coordinator of the Operative Plasterers' and Cement Masons' Job Corps Training Program, and Jerry Woods, executive director of the Illinois Ready Mixed Concrete Association. They agree that once the trowel makes a ringing sound, the surface is too hard for a broom finish.

The written documents we've seen don't clearly indicate the correct timing for a broom finish. For instance, one common specification for a trowel-and-fine-broom finish says to apply a trowel finish as specified, then immediately follow by slightly scarifying the surface with a fine broom. What's the meaning of "a trowel finish as specified"? The specification is ambiguous on this point because it says in a previous section that for a trowel finish you begin final troweling when the surface produces a ringing sound.

The section on brooming in the Portland Cement Association's *Cement Mason's Guide* (6th ed., 1995) can also be misinterpreted. It says to slightly roughen a hard, steel-troweled surface by drawing a damp, soft-bristled broom across the surface after troweling. The word "hard" in this description makes it sound as if several trowelings are done before the broom is used.

Peterson and Woods say that brooming must be done immediately after the second troweling. If you wait, says Peterson, the concrete may get too hard for the soft bristles to leave brush marks. And Woods reminds finishers to keep the trowel flat during the second troweling. Tilting the trowel leaves ripple marks that won't be taken out by brooming.

Q. *Why do manufacturers limit the content of air in concrete in floors that are going to receive a dry-shake metallic floor hardener?*

A. Master Builders Inc., a manufacturer of dry-shake metallic floor hardeners, recommends a 3% limit on entrained air in floors that will receive its product. When applying a dry-shake metallic floor hardener, it is necessary to have some moisture at the surface of the slab so the shake can be thoroughly worked in. This is more difficult with air-entrained concrete because of the air content's effect on bleeding. If thorough incorporation of the dry shake is not achieved, scaling can result.

Excessive air contents also can lead to other surface defects, including blistering. Blisters are caused by air and water pockets trapped beneath the dense, troweled surface of the slab. Because of the prolonged bleeding associated with high air content, finishers usually begin floating and troweling too soon, thinking that the bleeding has stopped. If the surface is sealed too early, further bleeding can lead to blisters.

Can you broom texture a hard-troweled surface?

Limit air content in floors to receive dry-shake hardeners

Steel troweling exterior air-entrained concrete

Q. *I've heard that it's not a good idea to use a steel-trowel finish for air-entrained exterior flatwork. However, I can't find any references that specifically recommend avoiding this practice.*

A. ACI 330R-92, "Guide for Design and Construction of Concrete Parking Lots," section 4.5.2, recommends leveling air-entrained concrete with a bull float or scraping straightedge immediately after strike-off, then dragging wet burlap over the surface as soon as the concrete has set enough to maintain a texture. The section further states: "The use of hand or power floats and trowels is not necessary and is not recommended, as their use may result in scaling." The most recent edition of the Portland Cement Association's *Cement Mason's Guide* contains the following statements:

- "Because of the danger of scaling or blistering, troweling of air-entrained exterior flatwork is not recommended.
- "Don't hard trowel air-entrained concrete surfaces that will be exposed to freezing and thawing, deicing agents or both."

Causes of spot retardation

Q. *What causes one part of a concrete slab pour to set slower than the rest of the pour? The amount of concrete involved is less than a truckload. The finishers have to wait longer for these spots to set up or settle for a slab that's wavy because some parts of the concrete are softer than other parts.*

A. Guy Detwiler of Material Service Corp. in Chicago says that spot retardation can often be traced to one of three causes:

- A wet spot in the subgrade, often caused by a low spot that collects water used to dampen the base course.
- On pumped concrete jobs, concrete that's been wetted in the pump hopper. The last portion of a truckload of concrete is often rocky or drier than the rest of the load. Water is sometimes added to improve pumpability, but the added water also means the concrete will take longer to lose enough water to be ready for finishing.
- Spilled soft drinks or coffee that contains sugar. Workers sometimes toss the remains of a cup or can into the fresh concrete. Sugar is a potent retarder, so the affected area won't set as fast.

Dick Gaynor of the National Ready Mixed Concrete Association says changes in the interaction between the cement and the water-reducing admixture may be the cause. To solve the problem, assuming the mix has worked in the past without causing problems, he advises the producer to: change cements, remove the water-reducer or retarder, or cut the admixture dosage in half. Gaynor also says water-reducing admixtures and retarding admixtures are more effective if the cement is wetted and the early hydration reactions have a few minutes to start before the admixture is introduced.

Preventing trowel burns on white floors

Q. *We're building a floor that calls for a white dry-shake to make it light-reflective. The specs also call for a hard troweled finish. We're concerned about burning the white surface. Is a stainless steel or plastic trowel better than an ordinary steel trowel? What else can we do to prevent burn marks?*

A. Trowel burns are caused by hard-troweling a surface so stiff that the trowel angle can be very high without causing chatter marks. Tilting the trowel increases pressure on the trowel edge and this densifies concrete at the surface. Darker concrete in the burned areas is simply concrete with a higher density and lower water-cement ratio.

According to Bob Gulyas of Master Builders, using a stainless steel trowel will prevent rust stains that could occur if you used an ordinary steel trowel. But it's still possible to burn a surface with stainless steel trowel blades if pressure on the blade is high enough.

We're not sure about the effect of using plastic trowels. Plastic is a softer material than steel and perhaps wouldn't have as great a tendency to burn a white surface.

Q. *We have installed hundreds of thousands of square feet of suspended floor slab on metal decking using lightweight concrete, and we've found it's next to impossible to get a smooth, level floor. All our work is done with power trowels, and when the slab is at its optimum for finishing, the concrete tends to roll slightly ahead of the power trowel as if it were a very stiff clay. This isn't noticeable during finishing but is certainly apparent in the completed slab, particularly as light passes over the surface at an angle. We have spent a lot of money on grinding and patching to produce surfaces that are suitable for floor finishes. Are there any solutions for this problem?*

Finishing lightweight concrete on elevated decks

A. Eldon Tipping of Structural Services Inc., Dallas, says this problem is related to the entrained air normally added to lightweight concrete to enhance workability and pumpability. The entrained air slows bleeding and can cause the surface to crust over before underlying concrete has set. Crusting is even more likely on elevated slabs because there aren't any natural wind breaks. Because the crusted surface looks ready for troweling, finishers get on it too soon and produce the wavy surface you describe.

Tipping offers several remedies. One is use of a spray-on monomolecular film that retards water evaporation from the surface. This product is available from admixture manufacturers. If finishers test a section to see whether it's ready for floating and troweling, and find it hasn't set enough, they should respray the test area.

If running water is available, another option is using foggers on the upwind side of the slab pour. Landscape nozzles allow you to dial in the desired fogging rate (from 1 to 5 gallons per minute). These nozzles can be mounted on a series of 3-foot-high vertical pipes connected by water hoses and will help to reduce the evaporation rate.

Finally, Tipping suggests trying pan floats instead of float blades to improve floor flatness (see *Concrete Construction*, May 1995, pp. 439-444). Using pan floats on ride-on, nonoverlapping power trowels can raise F-numbers 15 to 20 points. The pan floats can also be used on walk-behind power trowels, but the pan diameter should be less than 30 inches. Larger-diameter pans make the walk-behind machines too hard to handle because of the high torque.

Chapter 11

Formwork

Q. *The concrete subcontractor on our job is placing grade beams that are 8 inches wide and 40 inches deep. A nonchloride accelerator has been added to the mix, and the subcontractor wants to strip the forms four hours after concrete placement so workers can reset the forms for the next day's pour. Is there any problem with stripping forms this early?*

A. ACI 347R-94, "Guide to Formwork for Concrete," suggests that when the engineer/architect doesn't specify a minimum stripping strength, sides of beams can be stripped after 12 hours under ordinary conditions. This time represents the cumulative number of hours, not necessarily consecutive, during which air temperature around the concrete exceeds 50° F. However, the document also says this period can be reduced as approved by the engineer if high-early-strength concrete is used.

Because the concrete subcontractor is using an accelerator, it appears reasonable to strip the forms in four hours if this can be done without damaging the concrete surface. However, you should have the engineer approve this and recommend the proper time to backfill.

Q. *When renting or purchasing prefabricated column forms, we frequently see the note, "Brace according to industry standards," or similar wording. What are these industry standards, and who sets them up?*

A. It's hard to find a single "industry standard." The American Concrete Institute (ACI) "Guide to Formwork for Concrete" (ACI 347R-88) states the basic idea that braces should be designed to resist all foreseeable lateral loads such as seismic forces, wind, cable tensions, inclined supports, dumping of concrete, and starting and stopping of equipment. It recommends a minimum wind load of 15 pounds per square foot for design, but doesn't offer any specific suggestions for column form braces.

The actual number of braces needed may be affected by size or location of the column in the building. Is it more than one story high? Is it isolated or part of a long line of columns? Do beam, girder, or slab forms frame into the column form? The formwork handbook of one major construction company says simply "column forms must be braced in two directions," and this is about as close as we come to an industry standard for one-story columns.

In his book, *Formwork*, M.P. Hurst says column forms up to about 20 feet high must be supported on two adjacent sides so that they can be plumbed from two directions.

Another book, *Carpentry in Commercial Construction*, agrees that column forms must be plumbed and braced in two directions, and adds some specific recommendations: "Bracing is usually accomplished by nailing 2x4's near the top of the form and to a stake driven into the ground. These braces should make an angle of approximately 45 degrees with the horizontal

How soon can grade-beam forms be stripped?

Bracing for column forms

How much bracing is required for column forms? Bracing in two directions, as shown on these glass-fiber-reinforced column forms, is needed to plumb the forms. Sometimes local conditions call for more to maintain stability until placement of concrete is complete.

and should be placed on two adjacent sides...some contractors prefer to use the adjustable brace clamp produced by various manufacturers.... If there are a number of columns in a row, the end columns are braced in two directions with diagonal braces, and the central columns are held with horizontal spacers placed near the top to hold them in one direction. Diagonal braces are needed only in the other direction."

A formwork guide published by The Concrete Society in England (Ref. 4) says that column forms are "aligned, plumbed and stabilized with suitable props, preferably on all faces, to eliminate the possibility of twist" during concrete placement.

References

1. Hurst, M.P., *Formwork*, Construction Press, New York and London, 1983, page 189.

2. Badzinski, Stanley, Jr., *Carpentry in Commercial Construction*, Prentice Hall, Inc., Englewood Cliffs, New Jersey, 1974, pages 94 and 95.

3. ACI Committee 347, "Guide to Formwork for Concrete (ACI 347R-88)," American Concrete Institute, P.O. Box 9094, Farmington Hills, Mich. 48333, 1988, Section 2.2.3.

4. *Formwork: A Guide to Good Practice*, by a joint committee of The Concrete Society and The Institution of Structural Engineers, published by The Concrete Society, London, 1986, page 123.

Q. *What are the pros and cons of using redwood stay-in-place divider strips and bulkheads for flatwork?*

A. Stay-in-place wood bulkheads and divider strips produce neat joints that can be used to create a pattern of varying sizes of rectangles and squares, or to separate areas with different surface finishes. The strips also serve as contraction joints. This is especially useful with some exposed aggregate finishes for which hand-tooled joints aren't practical. For do-it-yourselfers who make small batches of concrete at the jobsite, stay-in-place forms break flatwork into smaller areas for better control of concrete placing and finishing. In many climates, a durable species of wood that's properly sealed will last many years in this application.

In warm, humid areas, however, even decay-resistant wood that's been sealed may rot. Even when the concrete is in excellent condition, it's hard to make an acceptable-looking repair. Also, load transfer between adjacent panels separated by wood strips is reduced because there's no aggregate interlock. Thus, the method may not be suitable for driveways or other surfaces that will be heavily loaded.

Stay-in-place divider strips and side forms for flatwork

Q. *When designing one-sided wall forms, what lateral pressures should I use?*

A. When designing one-sided wall forms, use the same lateral-pressure equations as for typical double-sided wall forming systems. The pressure exerted on the one-sided form does not change by placing concrete against an existing wall, soil bank, structure, or another form. Pressure is, however, transmitted through the one-sided form differently.

ACI Committee 347, "Guide to Formwork for Concrete," gives the following equations to calculate the lateral pressures on wall forms. The equations apply for concrete made with Type I cement, weighing 150 pounds per cubic foot (pcf), containing no pozzolans or admixtures, having a slump of 4 inches or less, and normal internal vibration to a depth of 4 feet or less. R = rate of placement, feet per hour; T = temperature of concrete in form, degrees Fahrenheit; p = lateral pressure, pounds per square foot; w = unit weight of fresh concrete, pcf; and h = depth of fluid or plastic concrete, feet.

For walls with a rate of placement less than 7 feet per hour:
$$p = 150 + 9{,}000 \ R/T$$
with a maximum of 2,000 psf, a minimum of 600 psf, but in no case greater than 150 h.

For walls with a rate of placement of 7 to 10 feet per hour:
$$p = 150 + 43{,}400/T + 2{,}800 \ R/T$$
with a maximum of 2,000 psf, a minimum of 600 psf, but in no case greater than 150 h.

If the previous placement rates or conditions do not apply: p = wh.

Blowouts, bulges, and out-of tolerance walls often occur with one-sided forms due to improper bracing. Because the ties used in wall forming can't be used for one-sided forms, the bracing alone must be able to handle all the lateral pressure.

A typical scenario on a wall form would be to place 3,000-pound ties on 2-foot centers. A comparable one-sided form would require heavy braces at close spacing to handle a load equal to the amount of load handled by these ties. The article "Blind-side and One-sided Forming" in the June 1992 issue of *Concrete Construction* shows details of such a system.

Lateral pressures on one-sided wall forms

Q. *We're helping out on a high-rise job, where the owners are concerned with speeding up the construction cycle. They want us to find some electrically heated forms to obtain higher early strengths in the concrete. Where can we find such a product?*

Electrically heated forms

A. There are several patents in this field, and successful work has been reported with such forms in both Europe and the United States, some of it dating back to the 1970s. Typically, electrical wires are embedded in the form material and temperatures are controlled by thermostat. The form must be substantial enough to distribute the heat evenly to the concrete. Both steel and fiberglass-reinforced plastic in combination with wood or plywood have been used for electrically heated formwork.

However, we haven't found a current source of such forms for cast-in-place concrete. As recently as 1988, electrically heated forms were offered at World of Concrete, but that manufacturer appears to be out of business. We asked Anthony Adonetti of Structural Contours Inc., who has bid on some of the work for which heated forms were once proposed. He believes that such forms would not be cost-effective today, and suggested alternative ways of accelerating the strength development of the concrete.

Heated enclosures are one answer; electrically heated pads laid on top of the slabs are another way to speed concrete curing. Effectiveness varies depending on the surrounding temperature. Another way to speed stripping time is to change the concrete mix so that it develops the necessary strength earlier in the work cycle. The time saved is frequently well worth the extra cost of the concrete. This approach helped speed work on the Riverfront III high-rise job reported by *Concrete Construction* in November 1991 (p. 783). The Riverfront contractors achieved their stripping strengths in three days, using column-mounted deck forms for about 70% of the job. The column-mounted forms delivered their load to the building's columns, greatly reducing loads on the newly stripped slabs below.

Forming curved walls

Q. *We were asked to build a curved concrete retaining wall for a homeowner. The curve of the wall has a radius of 4 feet. Can we produce this curve using job-built forms?*

A. Forming curved surfaces can be accomplished by using plywood, which can be bent to the desired radius. However, the plywood must be thick enough to support the load, or pressure, of wet concrete yet thin enough to bend properly.

Plywood can be bent two ways: the strong way and the weak way. Bending plywood so that the grain is perpendicular to the supporting studs is the strong way. Bending it so the grain runs parallel to the supports is the weak way. Though bending plywood the strong way provides more support, it's more difficult to achieve a tight radius.

To produce a bend with a radius of 4 feet, try using ⅜-inch-thick plywood with the grain running parallel to the supports. (North American plywood has the grain running the long direction of the panel.) Saw kerfing, or cutting notches partially through the plywood sheet, will allow you to use thicker sheets of plywood while meeting the required radius. For best results, provide smoothly curved supporting members. This can be done by cutting 2x stock to the proper radius. Remember to provide additional studs and place concrete in smaller lifts, since the sheathing is being used the weak way.

Can heat of hydration melt expanded polystyrene?

Q. *We're placing massive concrete walls using forms that have expanded polystyrene form liners. The wall sections are 25 feet high, 32 feet long and 12 to 24 inches thick. The concrete is made with 600 pounds of Type III cement per cubic yard. Typically we pour the concrete at 1 p.m. and start stripping forms at 8 a.m. the next day.*

The job started last January, and we had no problems until April, when the form liner started melting and sticking to the concrete, making form stripping difficult. The expanded polystyrene has a melting point of about 160° to 180° F. Does the concrete generate enough heat during hydration to melt the liner?

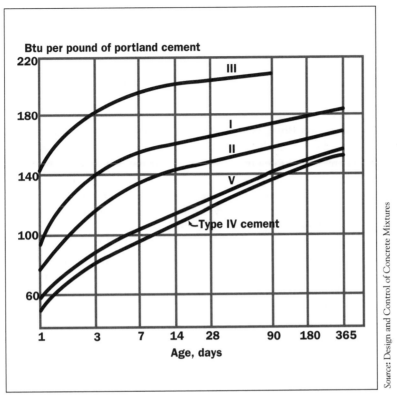

Btu per pound of portland cement

Typical heat-of-hydration curves for various types of cement.

Source: Design and Control of Concrete Mixtures

A. It's possible. In massive structures, the following equation can be used to estimate temperature rise when heat neither enters nor leaves the concrete:

T = CH/S

Where:

T = Temperature rise of the concrete in degrees Fahrenheit due to heat generation of cement when no external heat enters the concrete and no heat escapes

C = Proportion of cement in the concrete, by weight

H = Heat generation due to hydration of cement, in British thermal units per pound

S = Specific heat of concrete, in Btu per pound per degree Fahrenheit

Assuming that your concrete weighs 4,000 pounds per cubic yard, C = 600/4,000 = 0.15. From the graph below, heat of hydration of the Type III cement at one day is 140 Btu per pound. It would be a little less since you're starting to strip forms in less than 24 hours after concrete placement, but we'll use 140. Specific heat of concrete ranges from 0.20 to 0.28; we'll use the lower-end value, 0.20. Thus, T = 0.15(140)/0.20 = 105° F temperature rise. This assumes that because of the expanded polystyrene form liner, no heat was lost at the formed surface. If the concrete was placed at a temperature of 70° F, maximum temperature would be 105 + 70 = 175° F, which is within the 160° to 180° F melting point range you gave.

To test your hypothesis, embed a thermocouple near the formed surface before pouring the wall, and then measure the concrete temperature up to the time the forms are stripped.

Reference

Steven Kosmatka and William Panarese, *Design and Control of Concrete Mixtures*, 13th ed., Portland Cement Association, Skokie, Ill., 1994.

Magnets hold formwork blockouts

Q. *In the May 1990 issue of* Concrete Construction *(page 464) you showed magnets being used to anchor blockouts to metal forms. Is this a new idea? And where can people get the magnets?*

A. We first heard of this use of magnets while working on the article for our May issue, but it appears to be more common in Europe. C. J. Wilshire, writing about steel forms in the book *Formwork*, published in London, says "It is possible to get magnetic fixings—at a cost!"

Howard Gueldner and Bruce Harris of Symons Corporation's Steel Form Division in New Braunfels, Texas, told us that the magnetic blocks in the picture have pull-off load rating of 450 pounds, but the lateral force required to slide them along the form face may be only 40% of that. The blockout forms for that hotel job were custom-designed by Symons to accept several of the magnetic holding devices. Since each blockout form was used up to 80 times, the magnets saved a significant amount of cycle time. The manufacturer quotes a price of $51 for the B-450 magnet, which measures 1¼x2⅞x4¼ inches.

Gueldner pointed out that such magnets may lose up to 20% of their strength over time. One manufacturer reports that battery-powered electromagnets are available in sizes that would work with form panels. Manufacturers caution that holding power depends on steel thickness as well as on magnet strength.

Drilling through steel form panels to attach blockouts can be avoided by using permanent magnets to hold the blockout form in place during concrete placement. One magnet supplier reports that battery-powered electromagnets may also be able to do the same job.

A few of the magnet manufacturers, who may be able to offer more information, are listed below:

The Magnet Source
607 South Gilbert
Castle Rock, Colorado 80104
303-688-3966 (800-525-3536 outside Colorado)

Magnet Sales & Manufacturing Company
11248 Playa Court
Culver City, California 90230
213-391-7213 (800-421-6692 outside California)

Magnetic Division, Magni-Power Company
5511 Lincoln Way East
Wooster, Ohio 44691
330-264-3637 (800-221-6241 outside Ohio)

Q. *Soils in our area are stable enough for us to dig and form footings using the ground only. Is it permissible to build footings this way?*

A. Footings for residential and light commercial buildings are commonly formed this way in areas where the soils are cohesive enough to support a vertical face. However, Section 4.1.3 of "Specifications for Structural Concrete" (ACI 301-89) says that earth cuts shall not be used as forms for vertical surfaces unless required or permitted. Thus, on jobs with concrete specifications, you may need permission for earth forming from the engineer. You should also check local building codes.

The section on footings in "Formwork for Concrete" (ACI SP-4) says that when fabricated forms are omitted entirely and concrete is cast directly against the excavation, larger tolerances may apply. The book also suggests forming the top 4 inches when casting concrete directly against earth. This makes it easier to keep water and earth from washing into the excavation if it rains. Concrete foundation contractor Buck Bartley prefers to use a combination of lumber forms and earth forming where the soil will hold an edge. His crews set 2x4 side forms then dig the remaining depth of the footing. Bartley says if crews use no wood forms at all, it's much harder to build a level footing. And a level footing is the key to fast wall forming.

Q. *We placed a concrete wall for which the architect wanted an exceptionally smooth surface on the top 2 to 3 feet. We formed most of the wall with standard steel-faced forms but used hardboard at the top. When we removed the forms, the concrete surface in contact with the hardboard was soft and dusty. The rest of the wall looked fine. Are there chemicals in hardboard that retard concrete setting?*

A. Jerry Ford of Symons Corp. says that if surface appearance is critical, concrete shouldn't be cast against untreated hardboard because wood sugars in the product can leach out and retard concrete setting. Using form release agents on hardboard also can cause problems. The product data sheet for Magic Kote, a Symons form-release agent, advises avoiding use of the product on untreated Masonite. Ford says this is because the form-release agent softens the material.

Formless footings permitted?

Can hardboard retard surface setting?

Stripping wall forms during cold weather

Q. *We recently placed concrete walls in metal forms during a period when daily temperatures were around 50° F, but evening temperatures dropped to 25° F. The day after placing concrete, my foreman checked the concrete at the top of the form and found it was still green. How soon can we strip the wall forms?*

A. "Guide to Formwork for Concrete" (ACI 347R-94) offers some guidance on form stripping time. When air surrounding the concrete is at 50° F for a total of 12 hours (not necessarily consecutive hours), wall forms can usually be stripped. If high-early-strength concrete has been used, this 12-hour period may be reduced as approved by the engineer-architect. However, where wall forms also support formwork for slab or beam soffits, the removal times of that formwork should govern.

Consider several factors in deciding when to strip forms. After stripping, there should be no evidence of wall distortion or damage to the concrete surface caused by pry bars, wood shims, or other stripping tools. You must also protect the wall concrete from damage caused by freezing, both before and after stripping. You can do this by covering the forms with insulating blankets immediately after concrete placement, and covering the walls with such blankets after the forms have been stripped. Also, check specifications to see if additional moisture-retaining measures are required after the forms have been stripped.

Source for inflatable forms

Q. *We have a potential application for an inflatable form shaped like an 18-inch-thick pancake about 12 or 13 feet in diameter. However, we can't find a source for inflatable forms. Can you help?*

A. We know of one source:
Monolithic Constructors Inc., P.O.
Box 479, Italy, TX 76651 (214-483-7423; fax: 214.483.6662;
e-mail: mail@monolithicdome.com).

Cracking caused by cardboard void forms?

Q. *Last summer we poured a structural slab on grade over 4-inch-high cardboard void forms and a cardboard cover sheet. As shown on the detail, the slab is either 4 or 5 inches thick with #3 reinforcing bars at 16 inches on center in the 4-inch-thick slab section and #4 bars at 18 inches on center in the 5-inch-thick slab section. The problem: Severe irregular hairline cracking occurred over the entire slab, regardless of thickness. Generally, slab pour size was 3,000 to 4,000 square feet. The concrete had a 28-day design strength of 3500 psi and a design slump of 4 inches.*

Because the slab was placed during hot weather, we increased the 4-inch design slump to 5½ inches and added chipped ice at the plant to keep the delivered concrete temperature below 90° F. To cure the slab, we ponded each pour with ½ inch or more of water for five days.

Despite these precautions, all cracking began within 20 hours after the slab had been power troweled and then covered with water. The slab's structural integrity is fine, but the cracks are a cosmetic problem because they're exposed in the finished building.

This structural slab was placed over cardboard void forms because of severe expansive soil conditions. The second-floor structural slab, which was placed on plywood forms, has very few cracks; some areas have no cracking. This leads us to believe that placing concrete on cardboard forms caused the cracking.

Do you have any information on concrete placed over cardboard void forms? We want to reduce the cracking to a more acceptable level on future pours.

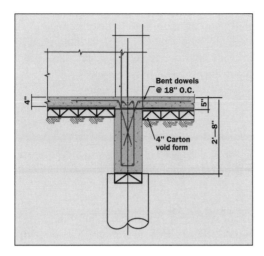

Bent dowels
@ 18" O.C.

4" Carton
void form

A. One consultant we contacted believes that the cracking might be related to heat buildup within the cardboard void form as a result of cement hydration. Venting the void forms might help reduce in-place concrete temperature. However, since the void forms are enclosed by grade beams, this might not be practical. It's also possible that flooding the slab with water caused a thermal gradient (cool on top, hot on the bottom) that produced the pattern cracking.

Q. *We're going to be placing two 60-foot-high concrete structures with a footprint that's 17x24 feet. They're to be placed in dewatered navigation locks and we have only 10 days to form and pour the concrete. The mix is very lean because engineers are concerned about heat buildup caused by cement hydration. We only need to hit 2500-psi compressive strength in 90 days. We'll be doing the pour when air temperatures are 30° to 40° F.*

We'll be using jump forms, but I'm concerned about form pressures and allowable stripping times. Because of the limited time, we need to make the lift height 20 feet. What kind of system should we use for forms this high? And when will the concrete be strong enough to support the anchorages used for the jump forming system?

**Form design
for relatively
low-strength
concrete placed
in cold weather**

A. You need to get help from your form manufacturer or from a forming consultant. With cold weather and a lean mix, form pressures will be high and you may have to design for pressures produced by a full liquid head of concrete. Lifts 20 feet high will require some pretty stout forms and bracing. You also may need some early-strength test results to decide when to strip the forms and when the concrete will be strong enough to support jump form anchorages. Get engineering help for these problems.

Q. *We manufacture precast concrete site furnishings with fiberglass molds. We currently use a sticky-back foam on the flanges. But we are still having leakage at the seam. The forms are keyed and bolted together at a 3-inch spacing. Is there a material we can use in place of the sticky tape that will not stain the concrete?*

**Solving seam
leakage on
fiberglass molds**

A. It sounds like the problem is more related to the mold's condition than to the type of seam filler. After a period of normal use, mold shells distort, warp or wear, causing poor fits at the shell seam. Overtightening mold bolts is the most common cause of shell warping.

One tip for avoiding shell warping is to loosely install all mold bolts before making them snug. One producer suggests adopting a mold-bolt tightening pattern, similar to tightening a tire's lug bolts, alternating from side to side. When mold shells are in good condition, many producers use paraffin or a petroleum-based jelly lubricant.

Worn or shrunk rubber liners also can cause leaky seams.

Long-line and short-line forming

Q. *Our precasting facility foresees the need of casting box girder segments at our facility. Unfortunately we do not have much space. How much room is required to produce bridge segments?*

A. Two methods of segment casting are available. Long-line or short-line forms can be used depending on the area available for the casting yard and the geometry of the bridge spans. In long-line casting, all segments are cast on a soffitt the full length of the cantilever or half length if the cantilever is symmetrical. All geometric control is accomplished while constructing the soffit, greatly simplifying control during segment production. A full soffit constructed for the long-line method, however, requires a large area, and the soffit might only be used once because it is difficult to accommodate variations for different bridge spans.

With short-line casting beds, the form is stationary while the individual segments move from the casting position to the match-casting position to storage and shipment. Advantages of short-line casting are much smaller space requirements, centralized production, adaptability to variations in bridge geometry, and the ability to reuse the forms many times. Casting bridge segments using a short-line bed requires accurate placement of the match-casting segment and post-casting geometry observations. Precise surveying skills and equipment are needed to measure elevations and alignments within 0.001-foot tolerances.

Determining minimum drafts on precast panels

Q. *We are casting a series of shallow-ribbed panels for a small office building. We have cast similar panels of this type for other buildings in the past without problems. This time, however, we are having*

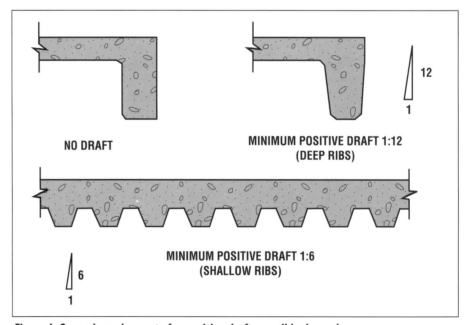

Figure 1. General requirements for positive drafts on ribbed panel.

problems stripping the panels from the molds. In several cases, we have had to disassemble the mold to free the panel. The panels are similar in size to those we have cast before. The only significant difference seems to be in the amount of positive draft on the ribs. The panels we have cast in the past had a positive draft of 1 inch in 6 inches (1:6) and 1 inch in 5 inches (1:5). On this project, however, the architect specified a positive draft of 1 inch in 7 inches (1:7). Could this be causing our problems, and how can we get the panels to strip easier?

A. Drafts on a panel allow the panel to be separated from the mold without disassembling the mold. According to the Precast/Prestressed Concrete Institute, generally, the minimum positive draft on the sides of openings and along the edges of panels for ease of stripping from the mold is 1 inch in 12 inches (1:12), with 1:8 preferred. This draft should be increased for narrow sections or delicate units where the suction between the unit and the mold becomes a major factor in both strength requirements and reinforcement of the unit. The draft should be increased to 1:6 for units pierced with many openings, for narrow ribbed panels, smooth concrete, and for very delicate units.

Drafts for ribbed panels should be related to depth, width, and spacing of the ribs (Figure 1). In deep sections, however, a 1:6 draft may result in an unattractive bulky design.

To minimize mold changes, discussion on draft should be initiated as early as possible after award of contract. However, since your project is already well underway, the only solution may be to look for a mold release agent that will help the panels strip easier. Ask for input from manufacturers and distributors of release agents.

Chapter 12

General Repair

Q. *We're doing some remodeling of a hospital building that includes installing new floor coverings. The floor slab is a lightweight concrete placed over metal deck forms. Evidently when the concrete was placed, the top surface froze and was patched. Then it was covered with a rubber flooring. We're removing the flooring now and the patches are coming up with it. What products can be used to patch the floor again and make a longer-lasting patch?*

A. If you are putting down new rubber covering, tile, or carpeting, you could probably use a cement-based flowable topping. This is an underlayment material that can be pumped into place and is nearly self-leveling. If any of the repaired floor won't be covered and will be exposed to hard-wheel traffic from carts, you might try a higher-strength flowable, cementitious underlayment or a polymer repair material. Contact one or more of the repair product manufacturers listed in *Concrete Construction*'s buyers' guide and ask for their recommendations. Remember to pay special attention to surface preparation when installing floor repairs.

Q. *We've read about using chain drags or striking concrete surfaces with a hammer to detect delaminations. Is there a standard procedure for using either of these methods?*

A. The equipment and procedures used to detect delaminations are described in ASTM D 4580, "Test Method for Measuring Delaminations in Concrete Bridge Decks by Sounding." Procedures for using both chain drags and an electro-mechanical sounding device are covered. You can order a copy of the test method from ASTM, 100 Barr Harbor Dr., West Conshohocken, PA 19428 (610-832-9585).

Q. *Does a patch applied to the vertical surface of a structural member (for example, a bridge pier) carry loads? I've heard that patches of this type serve no true structural purpose and are simply applied to protect the rebar.*

A. We spoke to consultant Russell Fling and he cited a number of factors that influence whether the patch will carry loads: location and depth of the patch, shrinkage and modulus of elasticity of the patching material, and thermal and structural movement of the column.

The patch will immediately carry the live loads applied to the structure for example, when a truck crosses the bridge. Unless the patch expands, however, the patch will not carry dead loads at first. When the structure spalls, the stresses in the structural member are redistributed to compensate for the loss of cross-sectional area. Patching the area will not immediately redistribute the dead loads. Whether the patch ever carries the dead loads of the structure (and how much load it carries) depends on the movement of the structure over time, shrinkage of the patch, and the depth of the patch. If the movement of the structural member due to creep and fatigue is greater than the amount of movement due to shrinkage of the patch, the patch will begin to carry loads. Higher modulus materials will carry more of the load than lower modulus materials.

How much load the patch will carry depends greatly on how deep the patch is. If the patch does not extend into the rebar cage, it will probably never carry much of the dead load and its purpose will primarily be to provide coverage for the rebar.

Effect of cracking on anchor capacity

Q. *When drilled anchors are used in concrete, isn't there a chance that a crack could form and pass through the anchor location, especially if the anchor is in a tension zone? What effect would a crack have on the anchor capacity?*

A. There is a chance that a crack could occur at the anchor location. And depending on crack width, anchor capacity could be significantly reduced.

German researchers (Ref. 1) studied the problem by forming cracks of differing widths through anchorage locations and then pulling the anchorage out. They compared capacities of undercut and expansion anchors in cracked and uncracked specimens.

For a crack width of 0.016 inch, failure load was reduced by 25% to 50%. Even at a crack width of 0.004 inch, a 20% to 25% reduction was measured. At crack widths beyond 0.016 inch, there was little further reduction in capacity over a wide range of crack widths.

British researcher A. W. Beeby (Ref. 2) believes, however, that cracks caused by loading of properly designed structures aren't likely to exceed 0.012 inch. He concludes that the factor of safety of 3 used in the United Kingdom for anchors includes allowances for possible strength reductions due to cracking. He also notes that there is no indication in the United Kingdom of unsatisfactory behavior of anchors as a result of cracking.

References

1. R. Eligehausen and R. Tewes, "Rationale for the UEAtc Draft Directives on Anchor Bolts Used in the Tensile Zone of Reinforced Concrete Members," University of Stuttgart, Report No. 1/29-88/4, December 1987.

2. A.W. Beeby, "Fixings in Cracked Concrete," Construction Industry Research and Information Association, Technical Note 136, 1990.

Using epoxy to rebond delaminations

Q. *How can epoxy injection be used to rebond delaminated topping slabs?*

A. Here's a step-by-step procedure to effectively rebond delaminated areas:

1. Map out the delaminated area. Use a chain or hammer sounding technique to locate the delamination. The delamination will sound "hollow" or dull when a chain is dragged across the surface or when a hammer is struck against the surface. Sweep off the surface and use a marker to outline the delamination. It's usually not necessary to to mark delaminations of less than 1 square foot.

2. Locate injection ports. Mark an X to locate a minimum of 4 ports at the outer periphery of the delamination, about 2 inches in from the outer periphery. For delaminations larger than 16 square feet, use an extra port for each additional 4 square feet.

3. Drill and set injection ports. Drill ½-inch-minimum-diameter port holes with a vacuum-attached swivel drill bit. Drill the holes deep enough to intersect the delaminated area. Use a fast-set epoxy to fix ports in place.

4. Clean delaminated area. Blow compressed air into drilled holes to force dust out of nearby holes. Limit air pressure to 50 psi to avoid causing further delamination. Some contractors prefer to use water to flush out dust. Either air or water can be used to make sure the drilled holes intersect the hollow separation. If air or water doesn't exit a nearby hole, redrill the holes.

5. Inject epoxy. Start injection, working from each port and moving across the delamination. Use an injection pressure of 25 to 35 psi. Increase the pressure to 45 psi for delaminations smaller than 2 square feet and to finish filling large areas. Hollow planes occuring along reinforcing bars may require additional resin, since cracks along the bars will fill with resin.

For additional information, see "How to Rebond Delaminations," *Concrete Repair Digest*, April/May 1994, p. 115.

Q. *As part of the renovation of an old steel mill, we have to make a large opening in a 3½-foot-thick concrete wall. The general contractor says that the concrete is too thick to be cut with a wall saw. How can we remove the concrete?*

Sawing options for bulk concrete removal

A. There are two other ways the concrete can be removed: diamond wire sawing or stitch drilling. Diamond wire sawing is feasible only if you have access to the back side of the wall. Otherwise, stitch drilling is the only option.

A diamond wire saw cuts with a steel wire that contains diamond-impregnated steel beads. The wire is threaded through holes drilled at the corners of the area to be removed, and the ends are joined by a steel connection sleeve. This continuous-loop diamond wire is mounted on a flywheel and directed to the cut by a series of idler wheels. The flywheel, driven by a hydraulic or electric motor, propels the wire. The wire loop is spun in the plane of the cut while being drawn through the concrete. The thickness of concrete that can be cut by diamond wire sawing is limited only by the power of the motor. However, you must have access to the back of the wall to loop the wire around the concrete. If you can't reach the back of the wall, stitch drilling is the only option.

Stitch drilling involves drilling overlapping bore holes through the concrete along the perimeter of the area to be removed. Its primary drawback is the risk of costly removal complications if the cutting depth exceeds the accuracy of the drilling equipment, so that uncut concrete remains between adjacent holes. Also, the opening created has very rough edges, so additional concrete removal may be required.

To locate a contractor with diamond wire sawing and stitch drilling capabilities, consider calling the Concrete Sawing & Drilling Association at 614-798-2252 for a list of contractors in your area.

Reference

ACI 546-96, *Concrete Repair Guide*, American Concrete Institute, 1996.

Q. *I'm familiar with impressed current cathodic protection, but I've recently heard about two other electrochemical concrete repair methods. One technique supposedly removes chloride ions from concrete and the other technique raises the pH of carbonated concrete. How do these techniques differ from cathodic protection?*

Electrochemical repair techniques

A. The techniques you are referring to, chloride ion removal (desalination) and realkalization, are very similar to cathodic protection in that the electrical hardware they use is the same. The major difference is that the current density for desalination and realkalization is about 50 to 500 times that used for cathodic protection. The other important difference is that desalination and realkalization are short-term treatments, whereas cathodic protection is normally intended to remain in operation for the life of the structure.

Realkalization differs from desalination in that the electrolyte that surrounds the anode contains an alkaline solution that is drawn into the concrete to increase its pH. Also, treatment times for realkalization are usually shorter than for desalination (days as opposed to weeks).

Removing high spots on a concrete floor

Q. *We're repairing an existing floor that will receive floor tile. All high spots that don't meet a ¼-inch-in-10-foot tolerance have to be removed. Is there a quick, low-cost way to do this?*

A. Lowell Brakey, a Kansas concrete contractor, has developed a slick way of doing this. His crew puts water on the floor to find the high spots. Then they grind the spots down using a walk-behind heavy-duty grinder that looks like a floor buffer but with diamond blades on the bottom.

Brakey says it's important to know immediately when to stop grinding because the blades cut so rapidly. To quickly alert the operator that it's time to stop, he mounted a laser receiver on the grinder. The crew first puts the grinder on a section of floor that's at the right elevation and adjusts the transmitter elevation until they get an on-grade indication from the receiver. Then they move to each high spot and grind until the on-grade indicator signals them to move on.

You also could grind down high spots with a terrazzo machine with Carborundum stones. But Brakey says the diamond grinders are much faster.

Cleaning the underside of rebars during repair

Q. *We do a lot of parking deck repair that involves removing delaminated concrete and concrete damaged by freezing and thawing. We remove concrete from beneath the rebars to make sure there's room for repair concrete to bond to the bars. But during the abrasive blasting operation that precedes concrete placement it's hard to clean the underside of the bars. Any suggestions for how to do this efficiently?*

A. Robert Kennerly of Sutton-Kennerly & Associates in Greensboro, North Carolina, has a method that seems to work pretty well. He suggests putting a thin steel plate beneath the rebars before abrasive blasting them. The abrasive bounces off the plate and hits the underside of the bars, thus cleaning both top and bottom of the bars.

Bonding agent needed for dam repair?

Q. *We're repairing a dam face that has suffered freeze-thaw deterioration to depths ranging from 6 inches to 2½ feet. The deteriorated concrete was removed by hydrodemolition. We're installing the formwork now and will begin pouring the repair concrete soon. The specifications require placing a polyvinyl acetate bonding agent onto the existing concrete before placing the repair concrete. Is this the right bonding agent to use? Is a bonding agent even required for this application?*

A. This is not a good application for a polyvinyl acetate bonding agent. According to ACI 546R-96, "Concrete Repair Guide," polyvinyl acetate should not be used in applications that may be exposed to moisture.

The Bureau of Reclamation's *Guide to Concrete Repair* recommends using an epoxy bonding agent for repairs that are less than six inches thick. According to the guide, thin repairs are subject to poor curing conditions caused by moisture evaporation and capillary absorption by the base concrete and, therefore, rarely develop acceptable bond strengths. Thicker repairs, however, do not require a bonding agent.

Q. *Most manufacturers of cementitious repair mortars recommend dampening the concrete substrate to a saturated, surface dry condition (SSD) before applying the mortar. It seems to me that the concrete substrate should be dry before applying the repair mortar to draw the cement paste into the concrete to develop a better bond. Is this reasonable?*

Dampening concrete before repair

A. We spoke to Rick Montani of Sika Corp., and he says that they've studied cores from patches where the repair mortar was applied over dry concrete. They found that, in many cases, the cement in the repair mortar suffered from a lack of water and incomplete hydration, resulting in weak and inconsistent bond strengths. He says that dampening the substrate to an SSD condition, then scrubbing the repair mortar into the substrate is the most practical way to ensure a strong and consistent bond.

However, there are opposing viewpoints. In the article "Mistakes, Misconceptions, and Controversial Issues Concerning Concrete and Concrete Repairs," *Concrete International*, November 1992, Ernest K. Schrader states that the ideal condition is for the substrate to be slightly drier than SSD. In "Secrets to a Successful Concrete Overlay," *Concrete Repair Digest*, February/March 1994, William F. Perenchio states that dry substrates result in higher bond strengths than wet or damp ones.

Q. *A specification requires us to use "dry pack" mortar to fill deep holes in a concrete wall. What is dry pack mortar and how is it installed?*

What is dry pack?

A. Dry pack mortar is a stiff sand-cement mortar that is typically used to repair small areas that are deeper than they are wide. According to the Bureau of Reclamation's *Guide to Concrete Repair*, dry pack mortar contains (by dry volume or weight) one part cement, 2½ parts sand, and enough water to produce a mortar that will just stick together while being molded into a ball with the hands. The ball should neither slump when placed on a flat surface, nor crumble due to lack of moisture.

Place dry pack mortar immediately after mixing it. Compact the mortar in the hole by striking a hardwood dowel or stick with a hammer. The sticks are usually about eight to 12 inches long and no more than one inch in diameter. Use a wooden stick instead of a metal one because metal tends to polish the surface of the mortar, making bonding less certain and filling less uniform. Place and pack the mortar in layers to a compacted thickness of about ⅜ inch. Direct the tamping at a slight angle toward the sides of the hole to ensure maximum compaction in these areas.

Overfill the hole slightly, then place the flat side of a hardwood piece against the hole and strike it several times with a hammer. If necessary, a few light strokes with a rag may improve its appearance.

Q. *We've been in the slabjacking business for about four years and have successfully completed many jobs. However, we sometimes have problems completely filling the void beneath the slab. What are some possible causes?*

Incompletely filled voids in undersealed floor

A. We spoke to John Meyers of American Concrete Raising Inc. and he offered some possible causes. The holes you are drilling to pump the grout through may be too big. Drilling large diameter holes increases the occurrence of "breakouts" in which concrete fragments break off the underside of the slab and settle in the void. If not detected and removed, these fragments can impede flow of grout to the void. Using a button bit rather than a star bit also helps to prevent breakouts.

Meyers also suggests that you begin pumping the grout under low pressure. High pressure removes water from the grout, creating a thicker mix. If you start pumping under high pressure, the mix may clog near the hole and prevent the grout from flowing into the remaining void. Pump the grout under low pressure until you get some resistance. Then gradually increase the pressure until you notice the slab lifting. Slab movement indicates that the void is full.

Sawcutting patch perimeters

Q. *Most specifications for partial-depth patches require the patch perimeter to be saw cut to a certain depth. I believe that this is good practice because it prevents featheredging of the patch material; however, I'm uncomfortable with the polished edge that is left after sawing. I believe the polished edge inhibits bonding of the repair material. Should this edge be prepared further?*

A. If abrasive blasting is also specified to remove rust from reinforcing steel or to remove loose concrete particles, then the edges of the sawcut can also be lightly blasted to provide a more roughened surface. If abrasive blasting is not part of patch preparation, consider making the sawcuts with a dry diamond blade rather than with a blade that requires water for cooling. Dry sawcutting does not leave a polished surface.

Repairing wet concrete surfaces

Q. *We are involved in a dam repair project that includes the repair of erosion-damaged concrete in the stilling basin. Most of the damaged areas are near the splash zone. We are considering repairing the concrete with an epoxy mortar to resist abrasion. However, the concrete surfaces are always wet and we're concerned that an epoxy will not achieve adequate bond to the concrete. Can an epoxy be used for this application?*

A. The ability of several cementitious and epoxy repair materials to bond to wet concrete surfaces was explored by the U.S. Army Corps of Engineers in technical report REMR-CS-25, "Spall Repair of Wet Concrete Surfaces," released in January 1990. Slant-shear bond testing was performed on 15 epoxies and seven cementitious products. To simulate wet surface conditions, the lower sections of the slant-shear specimens were stored in a moist room at 100% relative humidity until just before filling the upper half of the mold with repair material. The bond surface was examined just before filling the mold, and additional water was applied to the surface as necessary to obtain a glistening finish.

The subsequent slant-shear bond tests resulted in seven products achieving bond strengths of 1800 psi or greater. Of these, four were epoxies. These results suggest that certain moisture-tolerant epoxies can bond successfully to wet concrete surfaces.

Reapplying an epoxy bonding agent

Q. *We're placing a cementitious topping on an industrial floor, and the specifications require an epoxy bonding agent. The bonding agent must be wet or tacky when the topping is placed. After placing the bonding agent on a section of the floor, our mixer broke down and the epoxy dried. Can we reapply the same bonding agent over the section that has dried? If so, what sort of surface preparation is required?*

A. If you apply a second coat of the bonding agent soon after the first coat has dried, you shouldn't have a problem. Just make sure that the dried epoxy surface is clean. If, however, you wait more than a day or so, most epoxy bonding agents will develop an amine blush on the surface. The amine blush feels oily and will prevent bonding of the subsequent epoxy coat. In this case, you should clean the dried epoxy with a solvent such as acetone or toluene, then apply the second coat after the solvent has evaporated.

If the second epoxy coating dries before you are able to place the topping, you should seriously consider removing the bonding agent and starting over. Thick bonding agent layers are more likely to cause delamination of the topping due to differing thermal and structural movements between the epoxy and concrete.

Q. *We've been asked to remove a thin epoxy coating from a 20,000-square-foot warehouse floor. What methods can we use to remove the coating?*

A. Treating the floor with a chlorinated solvent (paint stripper) is an effective way to remove the coating without damaging the underlying concrete. Because chlorinated solvents are hazardous substances, they must be used with caution. Since the stripped coating will be a hazardous waste, you'll have to obtain a hazardous waste generation license, place the coating in labeled drums, and transport and properly dispose of the drums. After the floor is stripped, you must wash and neutralize it. Removing the coating with a paint stripper will not damage the floor surface, but if the existing coating is pigmented, it may leave a colored stain within concrete surface.

A second method to remove the coating is to shotblast the floor. However, shotblasting will take off the top surface layer of concrete along with the coating, leaving a roughened surface. This may be unacceptable if the owner does not plan to install a new coating to restore a smooth floor surface.

Another method is to use a recently introduced floor preparation machine that uses a belt sander and free flint abrasive to remove coatings. In addition to taking off the coating, this machine will abrade the concrete surface, but it will leave a smoother finish than a shotblaster.

Q. *We're making plans to repair a concrete beam that's been damaged by corrosion. The 100-year-old, nine-story building being repaired has a C-shaped footprint that extends through the first eight floors. At the ninth floor a beam spans the open gap in the C. The beam supports masonry walls for the ninth floor, some of the roof loads, and a masonry parapet on the roof. The beam spans 24 feet and is 5 feet deep and 18 inches wide. Much of the concrete has spalled off the underside of the reinforcing steel because of corrosion. We need to remove the bad concrete, clean up the steel, add new steel where needed, and apply a repair concrete. We think we'll have to remove concrete to a height of about an inch above the reinforcing steel for the full length of the beam.*

Here's the problem. Because the beam is 100 feet high, temporary shoring during repairs would be costly. However, removing all the concrete surrounding the steel may cause beam failure because of a loss of anchorage.

The beam ends rest on brick piers, with 20 inches of bearing length at each end. Concrete above the piers is sound, and a cover meter indicates that the tension steel extends through the bearing area. We think we could safely remove and replace small portions of the concrete by starting at each end and working toward the middle. The 20 inches appears to be enough anchorage to permit this approach. But we're also looking for other options. Is there a way to support the beam without using 100-foot-tall towers or shores?

Removing an epoxy coating from a warehouse floor

Repair method for corrosion-damaged beam

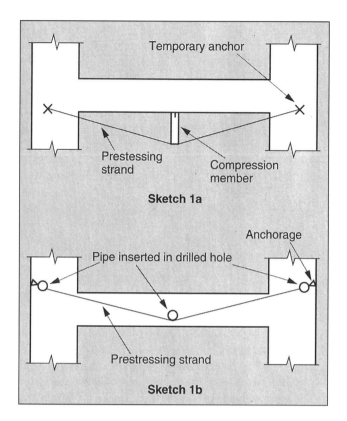

Sketch 1a

Temporary anchor

Prestessing strand

Compression member

Sketch 1b

Anchorage

Pipe inserted in drilled hole

Prestressing strand

A. You might try external post-tensioning. The method is sometimes used for permanent repairs but it could also possibly provide temporary support during repair. If there are suitable anchorage points at each end of the beam you could attach prestressing cables at the ends and place a compression member in the middle as shown in Sketch 1a above. Tensioning the cables applies an upward force at the beam midspan.

A second approach could be used if there isn't room for a compression member or saddle beneath the beam and if you have access to both sides of the beam. A hole is drilled above the tension steel at the center of the beam. A pipe is inserted in the hole, projecting from both sides of the beam. The prestressing cables would then pass as shown in Sketch 1b.

Ken Bondy of Seneca Construction Systems Inc. in Canoga Park, Calif., has repaired several structures using the external post-tensioning method. He says the method could be used for temporary support too. But he suggests that if it's a feasible approach for temporary use, it might be more cost-effective to consider a permanent repair using the method. At perhaps a lower cost than repairing the beam, a facade could be built to hide the cables and pipe or compression member while leaving them permanently in place.

Will surface preparation remove plastic sheathing from post-tensioning tendons?

Q. *We're going to apply an overlay to a deteriorated parking deck. In some areas, we have to remove concrete surrounding unbonded post-tensioning tendons. The tendons are encased in plastic sheathing. Is there any surface preparation method we can use that will remove the concrete but not the sheathing?*

A. We spoke to a few manufacturers of sandblasters and shotblasters and they said that using either of these methods will most likely damage, if not remove, the sheathing. However, Stan Smith, associate principal at J.R. Harris & Co., Denver, faced this challenge on a recent parking deck repair project. Mr. Smith said that a shotblasting demonstration performed on the project resulted in no apparent damage to the sheathing. But shotblasting removes the concrete surface slowly and is not an economical method if a significant amount of concrete needs to be removed.

Using chipping hammers, it's difficult to remove concrete from around the tendons without damaging the sheathing. Also, waterblasting the surface at pressures high enough to remove concrete will remove the sheathing as well. If you don't have to remove the concrete surface but just need to wash it, you may be able to pressure-wash the surface at pressures low enough to keep the sheathing intact. But Gerard McGuire, executive director of the Post-Tensioning Institute, advises against waterblasting over tendons encased in sheathing. By doing so, he says, you risk forcing water through any existing small holes or tears in the sheathing, increasing the risk of corrosion of the tendons.

One possible solution is to replace sheathing that is removed by surface preparation. Mr. Smith said that J. R. Harris & Co. is looking into the feasibility of applying a flexible, synthetic rubber coating to areas of the tendons where sheathing has been removed.

Q. *On most repair projects, we find that delaminated and deteriorated concrete often extends beyond the boundaries determined by chain dragging. For this reason, we usually prefer to chip out the deteriorated concrete first, then sawcut around the perimeter. But some owners and engineers object to this, saying that we should make the sawcuts first so our workers have a definitive point at which to stop chipping. Which method is correct?*

Sawcut or chip first?

A. This subject is often a source of great debate between owners, engineers, and contractors. The answer depends on which you believe to be a more accurate tool for identifying deteriorated concrete—the chain drag or chipping. Neither method is very scientific.

The chain drag relies on the operator detecting differences in the tone produced by the chain dragging across the surface. A hollow sound indicates a delamination; a ringing sound indicates good concrete. Some operators may not be able to hear the often subtle difference between the hollow and ringing sound.

To differentiate good concrete from bad when chipping, one rule of thumb is that aggregate particles in sound concrete will fracture when hit with a chipping hammer. In deteriorated concrete, they will simply separate from the cement matrix. However, in some lower-strength concrete, the aggregate may not fracture even though the concrete is sound. Once again, the method is subjective.

It's in the best interest of all parties to have all deteriorated concrete removed. Meet with the owner and engineer and point out areas where deterioration extends beyond the boundaries outlined in the condition survey. Then remove the remainder of the damaged concrete. If you can gain the trust of the owner and engineer, you may be able to get payment for additional work.

Q. *We're writing repair specifications for an industrial floor that needs to be undersealed with a cementitious grout to fill voids. Is there a standard undersealing specification that includes requirements for grout, hole patterns, equipment, and procedures?*

Specification for slab undersealing

A. We couldn't find a specification for undersealing industrial floors. But there are several good sources of information related to highway applications that would help you to write a specification. Here are three good sources:
- *Cementitious Grouts and Grouting*, Engineering Bulletin EB111.01T, Portland Cement Association, 5420 Old Orchard Rd., Skokie, IL 60076, 1990.

- *Pavement Stabilization by Undersealing*, ChemGrout Inc., P.O. Box 1140, LaGrange Park, IL 60526, 1985.
- *Cement Grout Subsealing and Slabjacking of Concrete Pavements*, IS212P, Portland Cement Association, 5420 Old Orchard Rd., Skokie, IL 60076, 1982.

Can slabs be lowered?

Q. *We're a residential slabjacking contractor and we often find that when one edge of a sidewalk slab settles, the opposite edge is raised. Raising the settled edge does not always make the opposite edge go down. Do you know of an economical way to lower a slab?*

A. When a slab has an elevated edge, most contractors find it more practical to raise the slab next to it rather than trying to lower the raised slab. Raising the adjacent slab eliminates the trip hazard, but it can look awkward. If this is unacceptable, most contractors opt to re-move and replace the raised slab.

One contractor we spoke to, however, does lower slabs when needed. He uses a steel jack with a leverage bar to raise the slab off the ground. He then scrapes out the excess subbase material and lowers the slab.

Installing floor covering over a patch

Q. *How long after a floor repair patch is made can a vinyl covering be applied?*

A. The general rule of thumb is to let concrete dry one month for each inch of slab thickness before applying a floor covering. Consult the floor covering manufacturer for more specif-ic concrete moisture content or relative humidity requirements and the manner in which each should be tested. For new construction, applying floor coverings 6 months after a 6-inch-thick slab has been placed may be acceptable. But waiting 2 months on a repair job until a 2-inch-thick partial-depth slab patch has dried sufficiently may be unacceptable. Generally, the own-er wants the repair to be done quickly, and that includes replacing the floor covering.

If the concrete hasn't dried to an acceptable moisture content or relative humidity level, you can continue waiting, add fans to accelerate drying, use desiccant drying, seal the concrete surface to reduce moisture emission, or use a floor covering that's suitable for a higher moisture content. Although it's usually not economical to dry out a floor during new construction, smaller patches sometimes lend themselves to easier and more economical accelerated drying.

A number of products are claimed by manufacturers to stop or limit moisture emission from concrete slabs. When using any type of sealer product, make a test section on your patch to determine the reduction in moisture emission rate. It's best to get the floor covering installer involved to make sure he will provide a warranty for the completed work. Also, check with the floor covering manufacturer to determine compatibility between the concrete slab sealer and the floor covering adhesive.

Making new concrete match old

Q. *We often place thin concrete overlays to repair surface defects such as small spalls and scaling. Because it hasn't weathered, the new concrete surface is often a different color than old adjacent con-crete. The new surface is usually lighter. How can we reduce the color difference?*

A. Some contractors use one or more muriatic acid washes to artificially weather concrete. Others suggest first etching the surface, then washing it with a mixture of carbon black and water. A third alternate is using a diluted black or brown concrete stain to darken the color. Make some test blocks with concrete from your normal supplier and experiment with several methods. Use the blocks as a color key for matching the older in-place concrete.

Q. *We're involved in the repair of a fire-damaged building. Much of the concrete is charred, but otherwise appears to be in sound condition. What effect do elevated temperatures have on concrete? Does the concrete need to be replaced after a fire?*

Fire-damaged concrete

A. The effects that fire has on the compressive strength of concrete are outlined in ASTM C 856 and are shown in the graph below. Petrographers usually can determine the temperature that the concrete reached in a fire by examining color changes in the aggregate.

According to Laura Powers-Couche, petrographer at Construction Technology Laboratories, the location within the concrete of the 700° F isotherm (a line connecting points of equal temperature) is important because, at this temperature, the reduction in compressive strength is about 50%. For practical reasons, however, the petrographer will locate the depth at which limestone aggregate is distinctly pink. This depth is the 570° F iso-therm and conservatively indicates concrete that is fire-damaged and probably will need replacing.

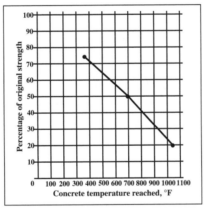

Reduction in strength of concrete containing siliceous gravel after heating, then cooling and testing.

Q. *My company is repairing the columns of a concrete building facade, and we have encountered areas where the reinforcing steel has corroded so badly that supplemental steel is needed. Do you have any advice for how and where the steel should be added?*

Supplemental steel for reinforced columns

A. We spoke to Jay Paul of Klein & Hoffman Inc., and he had several observations concerning the addition of supplemental steel to building columns.

When concrete delaminates or is removed from the column, the dead load is carried by the remainder of the column. Unless the column is unloaded by shoring it above the repair (which is very difficult and expensive), the repair will not carry any of the existing dead load. The designer must be aware of the redistribution of the load in a damaged column and determine if the column is overstressed. If it is, it might be necessary to relieve, or at least partially relieve, loads from the column before repair.

Ideally, supplemental vertical bars should be placed inside the column ties (rebar cage). However, this usually is not possible without cutting column ties or removing sound concrete from within the cage. Paul says that for fear of buckling vertical reinforcement he seldom allows a contractor to cut a tie. Even if the column ties are left intact, removing concrete from inside the cage could allow a vertical bar to buckle inward. It also further stresses the loaded column.

Whenever possible, Paul prefers to increase the size of the column and place supplemental bars outside the column ties. The bars can be supported laterally with stainless steel hairpin ties that are anchored with epoxy into the rebar cage.

If column ties are corroded, it's important to provide alternative lateral support to the vertical bars before removing any portion of the existing ties. Once again, hairpin ties can be installed that engage the vertical bars at the code-required spacing, but no less than the spacing of the existing ties. It is often necessary to build out columns to provide adequate cover over the supplemental ties.

Synthetic fibers vs. steel fibers

Q. *We're a wet-mix shotcrete repair contractor, and the engineer on an upcoming job doesn't want to use steel fibers as a replacement for 4x4-W2.0xW2.0 (4x4-8x8) welded wire mesh. Can polypropylene fibers be used instead?*

A. According to D.R. Morgan of AGRA Earth & Environmental Limited, you would have to add about 1% by volume (15.3 pounds per cubic yard) of collated, fibrillated polypropylene fibers to achieve the same residual load-carrying capacity as 4x4-W2.0xW2.0 welded wire mesh. The practical limit for most polypropylene fibers in shotcrete, however, is about 0.5% by volume (7.7 pounds per cubic yard). If you add more than this, the shotcrete becomes paste-starved and unworkable because too much cement paste is used to coat the fine fibers.

Morgan has, however, recently used synthetic polyolefin fibers to achieve residual load-carrying capacity similar to 4x4-W2.0xW2.0 welded wire mesh. The polyolefin fibers for shotcrete applications are 2½ times thicker than typical fibrillated polypropylene fibers. They can be added at a dosage rate of 15 to 20 pounds per cubic yard without adversely affecting the workability of the mix.

Undercutting corroded rebar

Q. *We were hired recently to oversee the repair of an elevated train platform. The contractor on the job has removed much of the delaminated concrete from the deck, but he doesn't plan to undercut the exposed rebar. He says that the underside of the rebar hasn't corroded so there's no reason to remove concrete from beneath the bars. He plans to wire brush or sandblast the rust off the top of the bars, then place the repair concrete. I would feel more comfortable if the rebar were undercut so that the repair concrete is mechanically anchored to the deck. Which method is correct?*

A. The International Concrete Repair Institute addresses this subject in their "Guide for Surface Preparation for the Repair of Deteriorated Concrete Resulting from Reinforcing Steel Corrosion." The guide recommends undercutting all exposed corroded rebar, regardless of whether the corrosion is believed to be limited to the top surface of the bar. The rebar should be undercut at least ¾ inch or ¼ inch more than the largest aggregate in the repair material, whichever is greater. If noncorroded rebar is exposed during concrete removal, these bars do not need to be undercut as long as their bond to the concrete has not been broken.

Corrosion— how much is too much?

Q. *Most specifications call for cleaning the rust off reinforcing steel before patching a deteriorated structure. How much cross-sectional area of the rebar can you afford to lose before you jeopardize the strength of the structure?*

A. The Technical Guidelines Committee of the International Association of Concrete Repair Specialists (IACRS) addresses this issue in the "Surface Preparation Guidelines for the Repair of Deteriorated Concrete Resulting from Reinforcing Steel Oxidation." According to IACRS, if a bar has lost more than 25% of its cross section, or if two or more consecutive parallel bars have lost more than 20% of their cross section, a structural engineer should be consulted to determine if repairs to the reinforcing steel are required. If repairs are required, the bar should be replaced or a new bar should be mechanically spliced to the old bar or placed parallel to and approximately ¾ inch from the old bar.

Q. *We do a lot of concrete restoration work and need to know what problems are associated with using epoxies for coating partial lengths of reinforcing steel. When corrosion of in-place rebar has started but hasn't yet significantly reduced bar diameter, our crews remove damaged concrete, clean the bar, and field coat it with epoxy. Then they patch with a cementitious repair material.*

I've heard that when only part of a bar is coated, corrosion is accelerated at each end of the coated area. And that even if the bar isn't coated, differences in the chloride content of the existing concrete and repair concrete cause faster corrosion in the unrepaired areas. Has this been proven in practice? Is it cause for concern?

A. For an answer, we contacted several sources including the Concrete Reinforcing Steel Institute (CRSI) and W.R. Grace & Co. There doesn't appear to be a well-documented answer to this question. Theory suggests that a corrosion-causing galvanic cell may be created if a rebar is partly covered by chloride-contaminated concrete and partly by uncontaminated concrete. Galvanic cells can also develop because of differences in oxygen and moisture concentrations between the existing and repair concretes. CRSI, however, says its field experience doesn't substantiate this. The organization has no data indicating that coating partial lengths of rebar has harmful effects or accelerates the corrosion process.

Neal Berke of W.R. Grace & Co. recommends that repair crews remove all concrete with a chloride content greater than 1.5 pounds of chloride ion per cubic yard of concrete. He cites a state highway department presentation on bridge decks where spot repairs had been made. After repairs were made, the decks continued to deteriorate in originally undamaged portions. Berke suggests advising the owner that replacing deteriorated concrete doesn't guarantee that other parts of the structure won't start to deteriorate. But he doesn't believe that using epoxy on partial lengths of rebar accelerates the corrosion process.

Q. *What is the repair procedure when extensive corrosion results in reinforcing bar section loss?*

A. Chip away concrete within the repair area to a depth sufficient to expose sound concrete over the entire repair area or to a minimum depth required by the patching material. Many engineers limit hammer size to 30 pounds or less to avoid fracturing underlying sound concrete. Concrete removal from around reinforcement is necessary when: a) rebar is rusted, b) more than half the rebar perimeter is exposed, c) concrete bond around the rebar is broken, and d) concrete around the rebar is unsound or honeycombed. A clear space of about 1 inch, sometimes referred to as a finger gap, is required under the reinforcement.

After the rebar is exposed, it can be cleaned and the surrounding concrete surface prepared by sandblasting. If wire brushing is used to clean the rebar, make sure to check the blind side of the rebar for effective cleaning. Once the rebar is cleaned, measure the remaining bar diameter. A common rule of thumb regarding the loss of rebar area (including deteriorated ties and stirrups as well as principal flexural reinforcement) is:

Does epoxy coating partial lengths of rebar cause faster corrosion of uncoated areas?

Dealing with rebar corrosion

Less than 10% reduction	No additional steel required
Between 10% and 20% reduction	Perform analysis to determine if additional steel is required
More than 20% reduction	Extra rebar required

Replacement or added steel must be carefully tied in the correct position (proper concrete cover and clearance) and overlapped at least 12 inches with the original bars. If the patch is too small to permit overlapping rebar, try welded or mechanical couplers or splices. Check with a structural engineer for advice on adding and replacing steel.

Floor isn't flat enough for racquetball courts

Q. *We placed the concrete for post-tensioned decks for a three-story health club. The specified tolerance for flatness was ⅛ inch in 10 feet. A racquetball court will be placed on wood sleepers spaced 8 inches apart on the floor, but we've found that the surface is out of spec. There are some high spots, up to ¼ inch, and some low spots that we've been asked to correct before they build the wood subfloor. What's the best way to do this?*

A. We suggest calling in a floor grinding specialist and working on cutting down the high spots instead of filling the low spots. You'll need to mark the high spots and check the elevation periodically as the grinding progresses. To get the names of grinding contractors in your area, call the International Grooving & Grinding Association at 615-449-8026. Once you've ground down the high spots, workers may be able to shim the low spots. Filling the low spots with a self-leveling underlayment is a possible solution, but isn't likely to work if wood sleepers will be fixed to the floor with powder-actuated fasteners.

Leaks between basement floor and walls

Q. *I recently received a call from a homeowner who says that, during heavy rains, water enters his basement between the floor and the wall. What are some possible causes of this? How do you correct this?*

A. Leaks between the basement floor and wall are usually the results of inadequate surface drainage or, if present, a poor drain-tile system. If water accumulates next to the structure, it can enter under a hydrostatic head at the location of the isolation joint between the wall foundation and floor. The preformed asphalt-impregnated fiber sheeting usually used to create this type of isolation joint is not an effective water seal. The joint allows the floor to move independently from the wall and foundation. The first line of defense against basement leakage is to drain surface water away from the structure. The second is to collect and drain the underground water away using a drain-tile system.

To drain water away from the surface, finished grade should fall off ½ to 1 inch per foot for at least 8 to 10 feet from the basement wall. Add fill material to offset reverse drainage caused by settlement or improper grading. On hillside sites, it may be necessary to construct cutoff drains on the uphill side to drain surface water around and away from the basement walls. Inspect downspouts and splash blocks to ensure they divert water at least 3 feet from the structure.

Drain-tile systems can be either inside or outside the footing (see figure). McCoy recommends evaluating the current system by opening the corners of the basement (for interior systems) or excavating (for exterior systems) to gain access to the drain tile. Then flush the drain tile with water to detect problems. McCoy says plugs, breaks, and low spots in the drain tile are the usual problems.

If a drain-tile system is not present and proper surface drainage doesn't stop the leak, install an interior drain-tile system. A properly installed system should eliminate this type of basement leakage and help ensure a dry basement. For tips on installing an interior drain-tile system, see *Concrete Repair Digest*, December 1992/January 1993, p. 261.

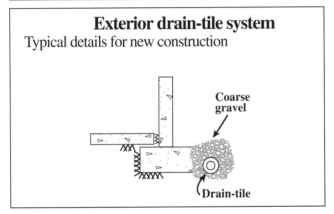

Interior drain-tile system
Install under basement floor to avoid problems associated with excavating outside of basement wall

Fill

Original grade

Isolation joint

Coarse gravel

Drain-tile (perforated pipe)

Exterior drain-tile system
Typical details for new construction

Coarse gravel

Drain-tile

Drain-tile systems prevent basement leaks by collecting and diverting underground water. Inspect the current system to locate plugs, breaks, or low spots. If a drain-tile system is not present and proper surface drainage doesn't stop the leak, install an interior drain-tile system.

Q. *We've been asked to correct an odor problem in a building that's more than 10 years old. It's a 5,000-square-foot building but the odor problem is only in one 15x15-foot office. We're sure the odor is coming from the floor. The concrete floor is discolored in that area and when the weather is damp there's a foul odor that smells like wallpaper sizing. The owner removed old rugs, scoured the floor, and put in new carpeting, but that didn't remove the odor. We taped down a square of plastic film on the bare concrete floor and got no moisture condensation under the film, so it doesn't appear that moisture is coming through the floor. How can we get rid of the odor?*

A. *Reader response:* We are a specialty flooring contractor and have for years been involved in solving floor problems. The one posed in your August-September issue is not unusual and can be solved.

The problem is moisture related even though the plastic-film test may have indicated otherwise. Sometimes, when a plastic film is taped to a floor where the temperatures of the subbase and the surface of the concrete are nearly identical, little or no condensation may develop.

Solving an odor problem

We suggest that the reader establish a substantial amount of air movement with a powerful fan and that the area be heated. This will help dry the concrete floor. If the problem is as explained, the odor should get more intense.

If the odor does get stronger, continue the heating and airing process for at least 24 hours. Then apply a penetrating vapor emission control product to the entire affected area (we use P-105, a potassium silicate product made by SINAK Corp., San Diego). Though the floor may be used as soon as the P-105 is dry to the touch, the treated area should be allowed a minimum of 14 days' curing time before carpet is applied.

Before installing the carpet, inspect the areas for residual odor. If not 100% satisfied with the results, the reader can apply an epoxy coating over the treated surface before the carpet is installed.

Richard E. Gregory, Allied Carpet, San Diego

Repairing a sagging, cracked balcony slab

Q. *We have been asked to fix a balcony slab that's sagged and cracked. It seems the top flexural reinforcement was placed near the bottom of the slab. Therefore, there's no tension steel and the balcony is barely able to support its own dead weight. What repair options are available? If possible, we'd like to avoid disturbing the tenants.*

A. You could install columns at the edge of the balcony or install brackets on the curtain wall to support the slab. Adding columns requires tearing up the ground next to the building to install foundations. Installing brackets requires cutting into the curtain wall. While these methods avoid disrupting the tenant, they do alter the appearance of the building. Superimpose columns and brackets on photos of the building to get an idea of how the building will look with the added members.

Another option is to cut into the slabs to bond new top bars in place. This requires a considerable amount of sawing and chipping but little repair material. The new top bars, however, would extend into the interior space. This repair method would disturb the tenants.

Instead of sawing and chipping to install new top bars, drill a hole from the edge of the slab inward. This method was used successfully on a 22-story building in Florida. On this project, holes were drilled 16 to 19 feet long and half-filled using polyester resin cartridges. Then a rebar was spun into place using drilling equipment. The contractor performed a trial repair to work the bugs out, then proceeded to drill holes and place rebar 2,100 times throughout the building.

Patching material for -10° F temperatures

Q. *We need to patch some core holes in a freezer room floor. But the freezer temperature is -10° F and the room can't be warmed to allow patching. Is there a product that can be used under these conditions?*

A. We know of one patching product that might work: Set 45, made by Master Builders Inc., Cleveland. However, some special precautions must be taken at this low temperature. It's best to mix the product at room temperature (about 70° F). Immediately before placing the patch, slightly heat the core hole with a propane torch (if this is permitted in the structure). Then cover the patched core hole with an insulating blanket and an electric heating pad.

Reader response: Reader Bruce Kahn of Great Wall Products, Waunakee, Wis., says that his company also has a product that will work at very cold temperatures. Called Speedset 60, the patching material reportedly has a compressive strength of 2000 psi after one hour, an ultimate strength of 10,000 psi, and meets ASTM C 928-92a, "Standard Specification for Packaged, Dry, Rapid-Hardening Cementitious Materials for Concrete Repairs."

Another reader—Jerry Loran of CTS Cement Mfg. Co., Cypress, Calif.—reports that his company's Rapid Set Nonshrink Grout and Rapid Set Mortar can hydrate at subfreezing temperatures. Both products have been used to make freezer-floor repairs at -10° F. Customers say they have been able to open up the repaired areas within 90 minutes.

Q. *We're considering the use of externally bonded, fiber-reinforced polymers for several concrete strengthening applications and would like to compile a list of sources. Who supplies these products?*

Suppliers of fiber-reinforced polymers

A. We are aware of nine suppliers of fiber-reinforced polymers:

C.C. Meyers Inc.
1401 S. Santa Fe Ave.
Compton, CA 90221
(310)223-2690
Product: Snap-Tite

External Reinforcement Inc.
10217 N. 87th Street
Scottsdale, AZ 85258
602-998-8616
Product: QuakeWrap

Fyfe Co.
6044 Cornerstone Ct. West, Suite C
San Diego, CA 92121
(619) 642-0694
Product: Tyfo S

Hardcore DuPont Composites
801 E. 6th St.
New Castle, DE 19720
(302) 427-9258
Product: HardShell

Hexcel Corp.
5794 W. Las Positas Blvd.
Pleasanton, CA 94588
(510) 847-9500
Product: Hex-3R Systems

Master Builders Inc.
23700 Chagrin Blvd.
Cleveland, OH 44122
(800) 628-9990
Product: MBrace

Sumitomo Corp. of America
1 California St., Suite 2300
San Francisco, CA 94111
(415) 984-3261
Product: Replark

Sika Corp.
P.O. Box 297
Lyndhurst, NJ 07071
(201) 933-8800
Product: CarboDur

XXsys Technologies Inc.
4619 Viewridge Ave.
San Diego, CA 92123
(619) 974-8200

The fibers, polymer resins and application methods for the systems can vary substantially. The reinforcing fibers are made of either carbon, E-glass or aramid, and the polymers are either epoxy, polyester or vinyl ester. With some systems, the fiber is supplied pre-impregnated with polymer resin; with others, all the resin is applied during installation. In addition, the fibers can be supplied in rolls, pultruded strips or thin shell for wrapping columns.

Q. *We're involved in a building facade repair project that requires partial-depth patches on the underside of several balconies. The patches are noticeable from the ground and the owner objects to their appearance. We've already installed elastomeric epoxy membranes to the top side of the balconies and*

What's a "breathable" coating?

are now considering coating the vertical and bottom sides. I know that in order to avoid trapping water in the concrete, I should use a coating that allows water vapor to pass through it, i.e., a "breathable" coating. What coatings are breathable? How is the breathability of a coating measured?

A. To measure the breathability of coatings, manufacturers commonly report the results of either of two standard test methods:
- ASTM E 96 Water Vapor Transmission of Materials
- ASTM D 1653 Water Vapor Permeability of Organic Coating Films

The test methods are similar but they use different test specimens.

In both tests, water is poured into a dish and the material to be tested is sealed over the top of the dish. The water is not in contact with the test specimen. The dish is then placed in a room having controlled temperature and humidity and weighed periodically. The weight loss of the dish indicates the amount of water vapor that has escaped through the test specimen.

The test specimen for E 96 can be any material as long as it has a uniform thickness of no more than 1¼ inches. When testing concrete coatings, the test specimen usually is a thin slab of coated concrete.

In D 1653, the recommended test specimen is a free coating film that is peeled off the substrate to which it was applied. But the test does allow the coating to be tested with film support such as filter paper or glass cloth if the cured coating is "too brittle or otherwise unsatisfactory for handling."

For both test methods, manufacturers usually list values for either water vapor transmission (WVT), expressed in grains/hour/square foot (7,000 grains = 1 pound), or water vapor permeance (WVP), expressed in perms. The higher the values, the more breathable the coating. The two values are not the same but are related:

$$WVP = WVT/\Delta p$$

where Δp is a measurement of the difference between vapor pressure in the dish and in the room.

There's no consensus in the coating industry as to what test results are needed to consider a coating breathable. Some manufacturers we spoke to offered suggestions on the definition of a "vapor barrier." Responses ranged from 0.1 to 1 perm. One manufacturer broke it down this way:

less than 0.75 perm—vapor barrier

between 0.75 and 3 perms—nonbreathable

greater than 3 perms—breathable

We spoke to technical representatives from several manufacturers about the coatings they recommend when water vapor permeance is important. The coatings included acrylic-based polymers, polymer-modified cementitious coatings, and water-based silicon coatings. Typical WVP values were between 8 and 12 perms.

New test method needed

We spoke to Fran Gale, director of technical services at Prosoco Inc., and chairperson of ASTM D01.47, a subcommittee on masonry treatments. Gale said that both test methods have their shortcomings when used to measure the water vapor permeance of concrete coatings.

E 96 does not contain provisions for how the coating is to be applied to the concrete substrate. Also, it measures not only the water vapor permeance of the coating, but that of the concrete substrate as well. The test method contains no instructions on how to prepare the concrete specimen or what its mix proportions should be.

D 1653 measures the permeance of the coating alone, but only film-forming coatings can be tested. If non-film-forming products such as penetrating sealers are tested, a film support must be used. Such support can have an effect on test results.

As a result, ASTM D01.47 has formed a task group to develop a test method specifically for measuring water vapor permeability of treated concrete and masonry substrates. The test method involves treating test specimens on five sides, then submerging treated and untreated specimens in water, removing them, and measuring their weight loss through the treated sides. It is hoped that the new test method will shed more light on the breathability of concrete coatings.

Chapter 13

Materials

Q. *I know that Type III cement is high-early strength cement, but what gives it a higher early strength than Type I cement?*

A. The main difference between Type I and III cements is fineness. Type III is ground finer. The greater fineness speeds cement hydration because there's more cement surface area that comes in contact with water. A faster hydration rate speeds strength development during the first seven days of curing. It also causes more heat to be released during the early curing period. Type III cements may also contain more tricalcium silicate and less dicalcium silicate.

Q. *Can using hard water as mixing water affect concrete properties?*

A. Increasing the hardness of mixing water can reduce the air content of air-entrained concrete. However, the effect can be counteracted by increasing the dosage of air-entraining agent.

Q. *We're designing a concrete structure in Egypt, and I received test results for the aggregate to be used in the concrete. The report lists chloride and sulfate content, but I don't know how to evaluate these numbers because ASTM C 33, "Standard Specification for Concrete Aggregates," doesn't give limits for them. Where can I get information about permissible levels of chlorides and sulfates in coarse and fine aggregates used in concrete?*

A. "Aggregates in Saudi Arabia: A Survey of Their Properties and Suitability for Concrete" was published in a 1987 issue of the RILEM publication *Materials and Structures*. It quotes the following maximum limits from the Ministry of Public Works and Housing's General Specifications for Building Construction:

NaCl maximum value = 0.1% for aggregates in reinforced concrete
SO_3 maximum value = 0.4%

Reader response: Your answer regarding permissible levels of chlorides and sulfates in aggregate might be OK for Saudi Arabia but certainly not for most of the United States. A limit on NaCl is meaningless: Does one measure sodium, or chloride, or both? Don't other chlorides count? What if both the coarse and fine aggregates contain chloride? (If so, you could have nearly 2 pounds of chloride per cubic yard, or 0.4% chloride by weight of cement, enough to reject the concrete even if all other components contain no chloride at all.)

Difference between Type I and III cements

Effect of hard water on concrete

Permissible levels of chlorides and sulfates in aggregate

For the SO_3 limit, 0.4% SO_3 by weight of concrete may be equivalent to about 3% by weight of cement, doubling the equivalent SO_3 in the portland cement and effectively surpassing any ASTM C 150 limit.

For U.S. designers, call on your friendly cement chemist for advice. Mine is: It depends on the type of chloride and sulfate in the aggregate, so have it analyzed.

William G. Hime, Erlin, Hime Associates, Northbrook, Ill.

Using gap-graded aggregate

Q. *What are gap-graded aggregates and how are they used in concrete construction?*

A. Aggregate is gap-graded when intermediate sizes are essentially absent from the gradation curve. Usual gap-graded mixes contain aggregate retained on a ¾- or 1¼-inch sieve, and particles passing the No. 4 sieve. Gap-graded mixes are used to obtain uniform textures for exposed-aggregate concrete and can also increase strength and reduce creep and shrinkage. Though the intermediate sizes usually can be omitted without making the mix unduly harsh or prone to segregation, choose the fine-aggregate percentage with care for gap-graded concrete. Use about 25% by volume with rounded aggregate and 35% with crushed material. Air entrainment usually is required to improve the workability of low-slump, gap-graded mixes.

Concrete drilling map

Q. *I'm looking for information about drilling holes into concrete. A number of tool manufacturers have alluded to a geological map of the United States that shows the varying hardness of concrete across the country. For instance, concrete in the upper northwest is supposedly very hard, requiring a rotary hammer drill to get the job done. In other areas you can get by with a hammer drill or even an ordinary drill.*

Does this map in fact exist? And if so, can you help me locate a copy?

A. You're probably referring to maps used by saw-blade manufacturers as an aid to picking the right saw blade. Such maps indicate areas in which aggregates are generally soft, medium or hard. We found one such map on a brochure from Blazer Diamond Products (800-245-5024) and another (pictured below) in a Target catalog (800-288-5040).

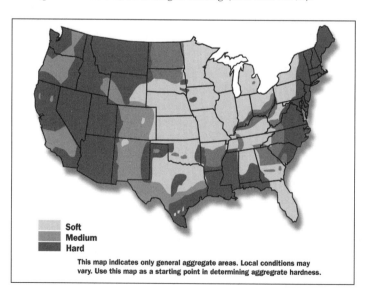

Soft
Medium
Hard

This map indicates only general aggregate areas. Local conditions may vary. Use this map as a starting point in determining aggregate hardness.

Q. *Is there a shelf life for portland cement, or will it last indefinitely?*

A. If kept dry, portland cement will retain its quality indefinitely. However, bagged cement that's stored for long periods in a dry atmosphere can develop what's called warehouse pack, a mechanical compaction that makes the cement lumpy. This can usually be corrected by rolling the bags on the floor to break up the lumps.

Shelf life for portland cement

Q. *We're going to be placing concrete slabs on grade in an area where the water soluble sulfate content of the soil is 0.4% (severe exposure). Is the use of Type V cement mandatory for this exposure?*

A. In selecting cements for sulfate resistance, a low C_3A content is the main consideration. For severe exposures, Type V cement with a maximum C_3A content of 5% is specified in Table 4.2.1 of the ACI Building Code (ACI 318-89). The code commentary also says that in certain areas the C_3A content of other available types such as Type III or Type I may be less than 5% and that these cements are usable in severe exposures.

Reducing permeability is the most effective way to increase concrete sulfate resistance and the simplest way to do this is to reduce water-cement ratio. A 1989 study showed that for concretes made with low (less than 0.40) or very high (greater than 0.60) water-cement ratios, C_3A content of the cement didn't have a great effect on sulfate resistance. However, C_3A content did affect sulfate resistance of concretes made with intermediate water-cement ratios (0.45 to 0.55).

Reference

David Stark, *Durability of Concrete in Sulfate-Rich Soils*, Portland Cement Association, R&D Serial No. 097, 1990.

Alternatives to using Type V cement

Q. *I once saw some rules of thumb about the effects of adding water to a concrete mix, but I can't find them.*

A. The basic rule of thumb is that adding 1 gallon of water per cubic yard increases the concrete slump by about an inch. Because of the more fluid mix, air content also is likely to increase by 0.5% to 1.0%. Concrete compressive strength decreases because the increased water-cement ratio produces a weaker paste and the increased air volume means that there is less solid material present. The compressive strength decrease could be 200 to 400 psi, considering both factors.

Adding water

Q. *When a Type I-II cement is specified, does it mean you use one cement that's a compromise between the two types, or does it mean the concrete supplier has the option of using either a Type I or a Type II cement?*

A. Cement manufacturers can make a single cement that meets the ASTM C 150 specification requirements for both Type I and Type II cements. It's called a Type I-II or Type I-and-II cement and isn't a compromise because all of the standard chemical and physical requirements for both types are met.

What is Type I-II cement?

If Type I-II cement is being used to reduce heat of hydration in a structure, rather than to produce moderate sulfate resistance, an optional C 150 chemical requirement should also be specified. This optional requirement limits the sum of tricalcium aluminate and tricalcium silicate percentages to 58% for the cement. All Type I-II cements may not meet the optional requirement.

Type I/II cement

Q. *Bagged cement labelled "Type I/II" is being sold in our area. Is this a new type of cement in ASTM C 150?*

A. ASTM C 150, "Standard Specification for Portland Cement," covers eight basic types of cement Types I through V, and air-entraining Types IA, IIA, and IIIA. For each cement type, limits may be set for chemical composition and physical properties. Type I cement is general purpose, and Type II is for moderate sulfate resistance and moderate heat of hydration. Type I cement does not have specified limits on several chemical compounds and has the same limits as Type II for some others. To achieve sulfate resistance and limit heat of hydration, Type II cement limits the content of tricalcium aluminate (C_3A) to 8%, and the sum of tricalcium silicate and tricalcium aluminate ($C_3S + C_3A$) to 58%. The physical requirements for Types I and II are almost the same, except that Type II is permitted to gain strength somewhat more slowly.

Type I/II portland cement satisfies requirements for both Type I and Type II. Strength requirements meet those for Type I, and composition requirements meet those for Type II. The dual-type cement can be used where either type is specified. It cuts down on the costs of producing and storing two cements. It's also helpful on projects where out-of-town designers specify Type I, but local practice is to use Type II. Type I/II can be used without requesting a substitution for the specified material.

Effect of concrete-making ingredients on chloride ion concentration in concrete

Q. *We're designing a reinforced concrete parking structure that will be exposed to deicing agents. The ACI 318 building code requirements permit a maximum of 0.15% water-soluble chloride ion by weight of cement for corrosion protection. What steps should we take to ensure that this limit isn't exceeded? Should we specify maximum chloride ion content for each of the concrete ingredients? Which of the ingredients contribute most to the chloride content?*

A. When the amount of chloride ion present in each ingredient in concrete is known, you can calculate the total amount present in the concrete. However, there's a difference between water-soluble chlorides and total (acid-soluble) chlorides in the mix. Not all of the total chlorides in concrete are water soluble. Some are chemically combined and don't contribute to corrosion. Richard Gaynor* of the National Ready Mixed Concrete Association estimates that ½ to ¾ of the total chlorides in concrete are soluble in water. The Building Code Commentary suggests that each ingredient of concrete be tested separately for total or acid-soluble chloride. Then if the calculated total chloride content of the concrete is below the limits given for water-soluble chlorides, further testing wouldn't be required. If the calculated total chloride content of the concrete is above the limit, you may recalculate using different ingredients or you may test the concrete for water-soluble chlorides as discussed in the Building Code Commentary.

You can calculate total chloride ion content of the concrete as shown in the example. Chloride contents of the separate materials are based on typical values suggested by Gaynor. The admixture is a nonchloride water-reducing product. Note that the total chloride ion content is less than 0.15%, so the concrete surely meets ACI 318 requirements that water-soluble chlorides be below 0.15%.

You'll also notice that the different concrete ingredients contribute significantly differing amounts of chlorides. If it is necessary to check alternative sources of materials, look first at the materials that contribute the most chloride ions. In the example, the coarse aggregate contributes by far the most total chloride. Even if the mixing water had 250 parts per million (ppm) of chloride ion, the total chloride ion content would only rise from 0.069% to 0.078%. The total chloride content of nonchloride water-reducing admixtures will generally be between 0.05% and 0.50% by weight of the admixture. The example demonstrates the effect of an admixture with a total chloride ion content of 0.05%. If a nonchloride water-reducing admixture containing 0.50% total chloride ion were used, the total chloride content of the concrete would only rise from 0.069% to 0.070%. Nonchloride water-reducing admixtures don't significantly contribute to chloride ion content of the concrete.

However, using straight calcium chloride or an admixture containing a large quantity of calcium chloride will increase the chloride content of the concrete. Adding 2% calcium chloride by weight of the cement in the example above would kick the total chloride ion content up to more than 1%. Using a water-reducing admixture that also contained calcium chloride to offset retardation would also increase total chlorides, although not nearly as dramatically. Gaynor reports that normal set admixtures made with chlorides might contain as much as 20% chloride ion by weight of the admixture. At that concentration in the example, total percent chloride by weight of the cement would almost double, going from 0.069% to 0.135%. The new value, however, is still below the ACI 318 limit of 0.15%.

To control chloride ion content in concrete, perform a calculation as shown in the example. Base your calculations on total-chloride-content testing of the fine and coarse aggregate. Also determine the chloride content of mixing water and admixtures to be used. For admixtures, dosage rate should be taken into account. Convert the dosage rate to pounds per cubic yard of concrete and multiply it times the chloride content of the admixture as shown in the example. Add the contribution from all the ingredients and you'll be able to tell whether or not there's

Sample calculation of total chloride ion (Cl⁻) content

Ingredient	Batch quantities	Total Cl⁻ by weight of material	Calculation	Total Cl⁻ in pounds
Cement	600 lb./cu. yd.	0.004%	0.00004 (600)	0.024
Sand	1,200 lb./cu. yd.	0.001%	0.00001 (1,200)	0.012
Coarse aggregate	1,800 lb./cu. yd.	0.020%	0.0002 (1,800)	0.360
Water	280 lb./cu. yd.	50 ppm	0.000050 (280)	0.014
Admixture	5 oz./100 lb. cement	0.05%	0.0005(5)(6)(9/128)*	0.001

Total Cl⁻, pounds per cubic yard: 0.411

Total Cl⁻, % by weight of cement $\dfrac{0.411}{600} \times 100 = 0.069$

*Assumes that 1 gallon of admixture weighs 9 pounds. There are 128 ounces in a gallon.

a problem in meeting the 0.15% requirement. It will usually be necessary to prohibit the use of straight calcium chloride. You also may have to place other limitations on ingredients, depending on the results of calculations.

*Gaynor, Richard, "Understanding Chloride Percentages," *Concrete International*, September 1985, page 26. American Concrete Institute, P.O. Box 9094, Farmington Hills, MI 48333; telephone 248-848-3800.

Sources of colorful aggregates for concrete

Q. *Where can we get some pure white and pure black pebbles for a special concrete mix? Our exposed aggregate renovation job calls for matching some existing concrete, and it looks like black and white were both used.*

A. In June 1988 (page 571) *Concrete Construction* published a list of 22 companies supplying a virtual rainbow of colors in crushed stone, gravel, and volcanic rock from numerous sources. Since then, we've heard from several other suppliers of decorative aggregate, so we're publishing an addition to the list here in Problem Clinic. We're also showing one of the photos sent in by Robert M. Lentz of Genstar Stone Products Company to demonstrate the versatility of architectural precast concrete surfaced with exposed aggregate.

Recently one supplier of special aggregates and sand, Fister/Warren, compiled and published a list of aggregates that have been or are being used for various types of architectural concrete—exposed aggregate precast, cast-in-place, terrazzo, and tilt-up. It includes such exotic names as Devil's Head Quartz, SanSaba Yellow Verona Marble, Bull Mountain Red Granite, and Royal Salmon Feldspar. To make it an easy reference guide, they arranged the list alphabetically three ways: by name, generically, and by color.

Genstar Stone Products Company

Architectural precast concrete for the façade of this mixed-use building in Shirlington, Virginia, contains white to light gray dolomitic limestone aggregate. Sandblasting of the panels exposed enough of the stone to enhance the building's color.

Fister/Warren also offers help in identifying duplicate names. For example, the same pink quartzite is named (by different suppliers) Tifton, Rosebud, Kinso, and Coral. If the specifier asks for an aggregate that is no longer available, they also offer to suggest the best substitute at a convenient location. To contact Fister Quarries Group Inc. (formerly Fister/Warren) write to 2777 Finley Road, Downers Grove, IL 60515; telephone 630-424-6200 (800-542-7393)

Editor's update: For information about sources of decorative and special-purpose aggregates, see the July 1997 issue of *The Concrete Producer.* Architectural precast concrete for the façade of this mixed-use building in Shirlington, Virginia, contains white to light gray dolomitic limestone aggregate. Sandblasting of the panels exposed enough of the stone to enhance the building's color.

Effect of fly ash on setting time

Q. *The sands in our area are mainly river sands that are deficient in very fine particles (material that passes the No. 50 sieve). Concrete made with these sands is hard to finish, especially when a steel trowel is required. Our ready-mixed concrete supplier recommends that we use a concrete with fly ash added to improve finishability. However, I've heard that fly ash retards the setting of concrete. Is it possible to add fly ash without affecting the concrete setting time?*

A. Some fly ashes chemically retard setting. And when fly ash is used in concrete as a cement replacement the concrete is likely to set more slowly than a straight cement mix would.

If fly ash is not being used as a cement replacement, but is added just to increase fines, it often won't retard setting. It simply serves as an additional fine aggregate and makes the concrete easier to finish, easier to pump, and less likely to bleed excessively. If concrete is air-entrained, you'll probably need more air-entraining agent to get the same air content as a concrete that doesn't contain fly ash.

Bad-smelling mixing water may not harm concrete

Q. *We're building an airport runway in Florida. Water from a local well is to be used as mixing water, but it has a strong sulfur smell like rotten eggs. Will whatever is causing the smell affect concrete properties? Should we be looking for a different water supply? If the water is left in a glass overnight, the smell goes away.*

A. The smell is hydrogen sulfide, a dissolved gas present in some well waters. Research Department Bulletin 119 from the Portland Cement Association says that quantities of hydrogen sulfide in industrial waters may vary from 0 to 15 parts per million (ppm). As little as 0.5 ppm can cause an odor. But the report states that it's hard to imagine even 15 ppm in mix water materially affecting the strength development of concrete.

If you need test results to verify that there isn't a problem, have a testing laboratory make two sets of mortar cubes (ASTM C 109). Use the questionable water in one set and distilled water or water of known quality in the other set. The water is suitable if seven-day strengths of cubes made with it are at least 90% of the strengths for cubes made with known-quality water. If effect on setting time is a concern, you might also have the testing laboratory run time-of-set tests (ASTM C 191) using the questionable water.

Algae in concrete mixing water

Q. *We've got a concrete paving contract in a mountainous region where the only water source is a lake with some algae growth. We've read that water containing algae is unsuitable for making concrete, because the algae reduces strength by interfering with cement hydration or by producing high air*

Mixing Water Organic Content, %	Air Content, %	28-Day Compressive Strength, psi
0	2.2	4830
0.03	2.6	4840
0.09	6.0	4040
0.15	7.9	3320
0.23	10.6	2470

Water filtration system can remove algae particles from concrete mixing water.

contents. *Is there a limit on the permissible amount of algae in concrete mixing water? If so, how would we test for the amount of algae present? And if too much is present, could we use copper sulfate to kill the algae before using the water?*

A. Laboratory tests (Ref. 1) are the probable basis for prohibiting the use of algae-infested water as mixing water for concrete. A researcher made an algae concentrate by sieving lake water containing algae, then freezing the concentrate and thawing it immediately before adding it to the mix water. Total organic matter in the concentrate was determined from loss on ignition at 650° C, and this value was converted to organic content of the mixing water, based on water content of the concretes tested. The researcher's results are shown in the table. Except at very low levels, increasing organic concentrations increased air content and decreased strength.

If no other source of organic material is in the water, a testing lab could probably determine algae content by inexpensive loss-on-ignition tests (about $30 per test). However, we caution against total reliance on such a test. Instead, have the lab test concrete made with the water, which will tell you immediately if there is an air-content or time-of-set problem. And 28-day cylinder tests on the same concrete will indicate the effect of algae on strength. If you need quicker results, consider accelerated strength tests.

We also advise against using water containing copper sulfate unless lab tests demonstrate no harmful effects. A broad survey of the effects of inorganic salts on strength of cement paste showed that copper chloride (in an amount chemically equivalent to a 2% calcium-chloride dosage) retarded set so much that no tests could be made at two and three days (Ref. 2). Depending on concentration, the sulfate ion might also lower the concrete's sulfate resistance.

You might consider buying a trailer-mounted water-filtration system similar to the one shown in the photo. This system has a flow rate up to 700 gallons per minute through a 6-inch-diameter filter that removes particles down to 0.004 inch in diameter. Semi-automatic cleaning with nylon brushes during the filtering process eliminates downtime during cleaning.

The manufacturer is Amiad Water System Technologies, 2220 Celsius Ave., Unit B, Oxnard, CA 93030 (805-988-3323).

References

1. Benjamin C. Doell, "Effect of Algae Infested Water on the Strength of Concrete," *Journal of the American Concrete Institute*, Dec. 1954, pp. 333-342.
2. Harold H. Steinour, "Concrete Mix Water How Impure Can It Be?" *Research Department Bulletin RX119*, Portland Cement Association, Skokie, Ill., 1960, p. 43.

Q. *How does water that's absorbed by aggregates affect concrete's water-cement ratio?*

A. Water-cement ratio is defined as the ratio of the amount of water, exclusive of that absorbed by the aggregates, to the amount of cement in concrete. In calculating the water-cement ratio, you don't include any water contained in the aggregate pores.

However, if air-dry absorptive aggregates are being batched, the amount of water added at the plant must be increased to maintain the desired concrete slump and water-cement ratio. If the plant operator doesn't make this batching adjustment, the concrete can lose slump as aggregates absorb water. This is more likely when aggregate absorption exceeds 1%.

Effect of absorbed water on water-cement ratio

Q. *We see many references recommending dilution ratios for muriatic acid that's used for efflorescence removal, acid etching, and other surface preparation applications. The references imply that there's a standard muriatic-acid solution, but our construction-supply distributor offers muriatic acid in several concentrations. What is muriatic acid, and is there a standard concentration?*

A. In the building trades, muriatic acid has become the accepted trade name for hydrochloric acid. Commercial-grade muriatic acid is a mixture of acid and water that is commonly 31.45% acid by mass (20° Baumé).

Percent by mass is the most precise way of specifying acid strength. You can make a 2% solution by adding 1 gram of 20° Baumé acid for every 15 grams of water. But most field solutions are mixed by volume. In inch-pound units, adding ½ pint of 20° Baumé acid to a gallon of water produces an approximately 2% solution by mass.

As you noted, however, you can also buy muriatic acid that contains less than 31.45% acid by weight. The concentration should be printed on the container or on the Material Safety Data Sheet.

Even when you know the concentration of the muriatic acid you buy, it still may be difficult to decide how much acid to add to water for the correct dilution. Instructions we've seen in magazine articles and other publications can be confusing. One says to use a 1% muriatic-acid solution. Another says to add ½ pint muriatic acid (without giving the acid concentration) to 1 gallon of water, and yet another tells you to use 25% more acid than the standard ratio recommended by the manufacturer. It's probably safe to assume that when the acid concentration isn't stated, 20° Baumé is implied.

Use caution when working with muriatic acid. Use it only with adequate ventilation, and wear proper protective equipment such as goggles, acid-proof gloves, and a respirator. When diluting the acid, always pour it slowly into water. Adding water to acid can cause splattering.

Also be aware that chlorides cause corrosion of metals embedded in concrete. William G. Hime of Erlin, Hime Associates, Northbrook, Ill., doesn't recommend using muriatic acid on concrete with less than 1 inch of clear cover over rebar or on masonry walls where joint reinforcement is present.

What is muriatic acid?

Strength development of Type I and Type III cements

Q. *Is there a general guideline or rule of thumb about how much more strength you get from Type III than Type I cement at various ages?*

A. At ordinary temperatures of about 70° F, you can expect Type III cement to develop about 190% of the Type I strength at one day, 150% at three days, 120% at seven days, and 110% at 28 days. At three months, the strength of Type I and Type III cements will be about the same. At 40° F, the advantage of Type III over Type I is more pronounced and lasts for a longer time.

Cement paste soundness

Q. *To what does the "soundness" of a cement paste refer?*

A. According to *Design and Control of Concrete Mixtures*, published by the Portland Cement Association, soundness refers to the ability of a hardened cement paste to retain its volume after setting. Lack of soundness or delayed destructive expansion is caused by excessive amounts of hard-burned free lime or magnesia. Most specifications for portland cement limit the magnesia (periclase) content and the autoclave expansion. Since the adoption of the autoclave-expansion test (ASTM C151) in 1943, there have been very few cases of abnormal expansion attributed to unsound cement.

What is ferrocement?

Q. *We recently read about a project in which a tunnel was lined with ferrocement. We have heard of ferrocement before, but what exactly is it and on what types of projects is it commonly used?*

A. Ferrocement is a special type of reinforced concrete composed of closely spaced layers of continuous relatively thin metallic or nonmetallic mesh or wire embedded in mortar. It is constructed by hand plastering, shotcreting, laminating (forcing the mesh into fresh mortar), or a combination of these methods.

The mortar mixture generally has a sand-cement ratio of 1.5 to 2.5 and a water-cement ratio of 0.35 to 0.5. Reinforcement takes up about 5% to 6% of the ferrocement volume. Fibers and admixtures may also be used to improve the mortar quality. Polymers or cement-based coatings are often applied to the finished surface to reduce porosity.

Ferrocement is considered easy to produce in a variety of shapes and sizes; however, it is labor intensive. Ferrocement is used to construct shell roofs, swimming pools, tunnel linings, silos, tanks, prefabricated houses, barges, boats, sculptures, and thin panels or sections usually less than 1 inch thick. *Source: Portland Cement Association.*

Should fly ash be eliminated in an oversanded mix?

Q. *A swimming pool contractor who is a customer of ours wants an oversanded mix to use in shotcreting. We have suggested one of our oversanded mixes, which contains 730 pounds of cement, 180 pounds of fly ash, 700 pounds of ½-inch stone, and 1,800 pounds of sand. We sometimes supply this mix at a 6-inch slump to other customers who use it with a small concrete pump. This customer is afraid that the finish coat, which contains marble dust, will not bond properly to shotcrete containing fly ash. Should we reapportion the mix to eliminate the fly ash?*

A. There should be no problem with your mix. However, you might caution the contractor to keep the shotcrete moist so that the surface cures well and also to keep the surface clean. The contractor can accomplish this by keeping the surface covered with a plastic sheet.

Q. *Since aggregate occupies a majority of the volume of concrete, it must play a major role in strength. How will a change in coarse aggregate size affect the compressive strength? Should large- or small-diameter aggregates be used in the production of high-strength concrete?*

A. The strength of concrete depends on the strength of the paste, the coarse aggregate, and the paste-to-aggregate bond. Generally, the strength of normalweight aggregate is much greater than that of the cement paste, and is therefore, less important. Although the water-cement ratio and its effect on the strength of the paste is the most important factor affecting concrete strength, the properties of the aggregates cannot be ignored. Most important are the shape, texture, and size of the aggregate. The aggregate strength may limit concrete strength when lightweight aggregates are being used or when high-strength concrete is being produced. In normal strength ranges, with the same cement content and consistency, concrete mixtures containing larger aggregate particles require less mixing water than those containing smaller aggregate. This leads to a lower water-cement ratio and higher strengths. However, the larger aggregates tend to reduce the bond strength between the aggregate and the paste. The net increase in strength from using larger aggregates is slight. For each source of aggregate and concrete strength level, there is an optimum size aggregate that will yield the highest compressive strength per pound of cement. Trial batches should be made with various aggregate sizes and cement contents to find the optimum size. Many studies have shown that, although ¾- to 1-inch aggregate sizes have been used successfully, ⅜- to ½-inch maximum size aggregates appear to give optimum strength in high-strength mixes.

References

1. Mindess, Sidney and J. Francis Young, *Concrete*, Prentice-Hall, Englewood Cliffs, New Jersey, 1981, pp. 124, 344-347, 395-396.
2. *Design and Control of Concrete Mixtures*, Portland Cement Association, 1990, pp. 184-185.

Q. *Throughout the year, I have bought cement from three different cement producers. I know that there are slight color differences, but I've noticed other differences, such as water demand and slump loss. When I change cement during the winter, I may get complaints about setting time differences. Because I only buy Type I cement, shouldn't all the cement conform to the same standard?*

A. ASTM C 150 specifies performance levels for physical and chemical properties of cement. The problem is that range of performance permits several differences to exist within a type of cement.

For example, the initial setting time for a Type I cement as determined by the Vicat test is "not less than 60 minutes." Type I cements commonly vary up to 3 hours for initial set.

You certainly may see differences in the performances of cement between brands. Worse yet, fly ash and chemical admixtures work differently with various brands. Your best bet is to pay attention to how they perform in your concrete and use the ones that work the best.

It may help to talk to your cement salesperson about the problem. He or she may be able to offer you advice on how to solve the problem with varying performance levels.

Q. *We have some cement that has been stored in a silo for about one year. It still flows freely and is not lumpy, although a 1- to 2-inch-thick crust has formed over the surface. Is this cement still good? Also, are there any simple tests that can be made to determine if cement is still good?*

Coarse aggregates' effect on strength

Differences in cement

Cement stored for a year

A. As long as the cement in your silo has not been exposed to moisture, it is probably still good. Portland cement is a moisture-sensitive material; if kept dry, it will retain its quality indefinitely.

Cement stored in contact with damp air or moisture will prehydrate. Concrete made with such cement sets more slowly and has less strength than concrete made with cement that has been kept dry. Cement exposed to air will absorb moisture slowly, which will cause a crusty top to form. This crust indicates that the cement has started hydrating. Even if this occurs, however, most of the cement is probably still undamaged.

When used, cement should be free-flowing and free of lumps. If lumps do not break up easily, the cement should be tested before it is used. If you doubt the quality of your cement, perform standard strength tests or loss on ignition (LOI) tests. You can determine loss-on-ignition of cement (following ASTM C 114) by heating a cement sample of known weight to 900° to 1,000° C until a constant weight is obtained. Then you determine weight of loss of the sample. A high loss on ignition indicates the cement has started to hydrate or carbonate, which may be caused by improper and prolonged storage or adulteration during transport and transfer. Avoid using cement that has a high LOI.

Ordinarily, cement does not remain in storage long, but it can be stored for long periods without deterioration. Above all, cement should be stored at as low a relative humidity as possible. Bulk cement should be stored in weathertight concrete or steel bins or silos. Dry, low-pressure aeration or vibration should be used in bins or silos to help cement flow better and avoid bridging.

Small aggregate reduces slump

Q. *I know that slump decreases if I increase the amount of fine aggregate in a mix and don't adjust the water content, but it doesn't make sense to me. It seems that a sandy mix ought to slump more than a mix with more course aggregate. Why does adding more sand reduce slump?*

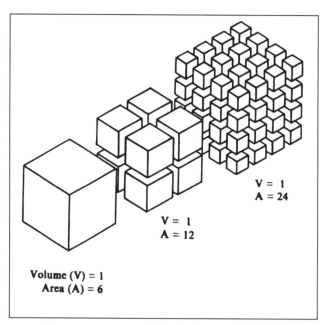

V = 1
A = 24

V = 1
A = 12

Volume (V) = 1
Area (A) = 6

Surface area increases rapidly as aggregate particle size decreases. A 1-inch cubical particle has only 6 square inches of surface area. Cutting this cube into 512 eighth-inch cubical particles doesn't change the weight but increases the surface area to 48 square inches.

A. In a workable concrete, each aggregate particle, from large to small, is coated with cement paste. The thicker the coating and the more watery the paste, the higher the slump. This is just another way of saying that you can increase slump by adding water and cement or by adding water alone. The second option, however, reduces concrete quality.

Small aggregates have a greater surface area for a given weight than larger aggregates (see figure). So if you replace large particles with many smaller ones and don't adjust the paste content, you'll get a thinner coating of paste and a lower slump.

Because of this surface-area effect, pea-gravel concretes, which require a higher sand percentage, also require a higher water content to produce a given slump. That's why concretes with a large maximum aggregate size are more economical (requiring less water and cement) than concretes made with smaller maximum size aggregate.

Segregation based on aggregate source

Q. *We have a problem with concrete segregation that appears to be related to the coarse aggregate we are using. When we use crushed stone from one source the concrete segregates in the chute of the ready mix truck. When we use stone from another source, we get no segregation. Both aggregates meet ASTM C 33 grading specifications. I would like to find some way to use the stone that causes segregation because it costs less and the source is closer to our plant. The concrete is air-entrained and we use the same sand in both mixes.*

A. You did not mention anything about mix proportions. The same weight ratio of coarse to fine aggregate may not work. ASTM C 33 grading limits are sufficiently broad to allow significant difference in gradings of the two coarse aggregates. The dry-rodded unit weight (DRUW) of the coarse aggregates should be determined. For coarse aggregates with similar specific gravities and nominal maximum size, a lower DRUW indicates a higher void content, in which case a higher sand content will be necessary. Follow ACI 211.1 procedures for proportioning the mixture.

Segregation is increased if the sand is deficient in fines or the specific gravity of the coarse aggregate is high. In flowing concrete prone to segregation, the material passing the No. 50 sieve, that is cement and sand fines, should be increased to exceed 750 pounds per cubic yard. Increasing the fines by using fly ash or additional cement should help. Another possibility is to change the grading of the problem aggregate so that it resembles that of the better aggregate. Replacing some of the problem aggregate with an intermediate size can improve the total grading and reduce segregation.

In general, coarse aggregate angularity and absorption will have an effect on water demand and, hence, slump. Lean mixtures at a high slump will be more susceptible to segregation. Experimenting with laboratory-size trial mixes should lead to an easy and inexpensive solution.

Reader response: Regarding your October Production Troubleshooting answer on variations of grading of aggregate: Changes in ASTM C 33 and ACI 301 were published in 1993. The net effect of the changes was to require that attention be given to the combined aggregate grading and not just that in stockpiles.

The aggregate causing the segregation problems may be cheaper because the supplier is taking advantage of the broad specifications allowed for concrete. The supplier probably can make more money by taking some of the -½ inch aggregate and selling it for asphalt or other purposes. The supplier cannot be faulted because that is the way low price is often achieved and aggregate grading can still be well within the standards the supplier agrees to meet.

The use of dry-rodded unit weight as a means to recognize the problem is true. However, from our experience in the field, the new mixture would either be too rocky or too sandy. Both could cause placing, finishing, and durability problems.

The only effective way to produce consistent concrete at the least cost is to blend three aggregates. The third aggregate would be a "blend size" as provided in ASTM C 33-93. The ideal blend for most parts of the country is graded from ½ inch to No. 50. It is too coarse to be defined as sand and too fine to be pea gravel. This aggregate is known by different names, including buckshot, bird's-eye, squeegee, and chat. It should be either natural, rounded, or cubically crushed. Elongated slivers cause more problems. With such a blend size, a producer can make adjustments as materials vary to produce a consistent combined grading.

This would allow production to meet the alternate requirement of the new ACI 301. That specification provides limits on maximum and minimum amounts retained on each sieve for the combined grading.

James M. Shilstone Sr., The Shilstone Cos. Inc., Dallas

Suppliers of ultrafine cement

Q. *As part of a dam repair project, we're installing grout curtains beneath the structure. On previous soil permeation projects, we've used sodium silicate as the grouting material. But we'd like to achieve a longer service life for the curtain and plan to use a cementitious grout. We've heard that ultrafine cements provide good performance for these applications. Who sells ultrafine cements?*

A. Ultrafine cements are available from the following sources:

De Neef Construction Chemicals Inc.
P.O. Box 1219
Waller, TX 77484
(800) 732-0166
Product: MC-500

Lehigh Portland Cement Co.
7660 Imperial Way
Allentown, PA 18195
(800) 523-5488
Product: Microcem

Master Builders Inc.
23700 Chagrin Blvd.
Cleveland, OH 44122
(800) 628-9990
Product: Rheocem

Fosroc Inc.
150 Carley Ct.
Georgetown, KY 40324
(800) 645-1258
Product: 1092 Ultracem

Surecrete Inc.
155 NE 100th St., Suite 300
Seattle, WA 98125
(206) 523-1233
Product: Nittetsu Super-Fine Cement

Blue Circle Cement Inc.
2 Parkway Center
1800 Parkway Place, Suite 1200
Marietta, GA 30067
(800) 282-6350
Product: Microcem 900

Nailable concrete

Q. *We had an inquiry for nailable concrete. We recall that sawdust concrete was promoted for this purpose at one time. Is there anything new on this subject?*

A. At least 45 years ago, sawdust concrete was indeed promoted for its nailability. Perlite concrete, vermiculite concrete, cellular concrete, and concrete made with foamed polystyrene beads, which in this country are usually made in a density class of about 20 to 50 pounds per cubic foot, all will accept nails. The same concretes made at higher densities will undoubtedly also do so. We know of no standard, however, of testing the nailability of concrete, nor any specifications for nailability.

In addition, don't forget that normal-weight concretes can be nailed with powder-actuated tools. The method apparently can be used successfully with concrete in the compressive strength range of about 1400 to 5000 psi.

Q. *What effect does aggregate type and size have on the drying shrinkage of concrete? What types of aggregates are least prone to shrinkage?*

A. Coarse and fine aggregates, which normally occupy 65% to 75% of the total concrete volume, have a significant effect on the drying shrinkage of concrete. According to ACI 224, "Control of Cracking in Concrete Structures," the higher the stiffness or modulus of elasticity of an aggregate, the more effective it is in reducing the shrinkage of concrete. The absorption of the aggregate, which is a measure of porosity, influences its modulus or compressibility. A low modulus is usually associated with high absorption.

Quartz, limestone, dolomite, granite, feldspar, and some basalts are generally classified as low-shrinkage producing aggregates. High-shrinkage concretes often contain sandstone, slate, hornblende, and some types of basalts. Some shrinkage values are shown in the table.

Maximum aggregate size also has a significant effect on drying shrinkage. Using large aggregate sizes allows for a lower water content of the concrete and is more effective in resisting shrinkage of the cement paste.

Effect of aggregate type on concrete shrinkage

Aggregate	Specific gravity	Absorption, percent	One-year shrinkage, percent
Sandstone	2.47	5.0	0.116
Slate	2.75	1.3	0.068
Granite	2.67	0.8	0.047
Limestone	2.74	0.2	0.041
Quartz	2.66	0.3	0.032

Source: American Concrete Institute

Q. *We make our own dry repair mixture for use on small patching jobs, but the sands in our area are coarse, making the repair mixture hard to finish. It also bleeds a lot. Someone recommended that I add fly ash to improve finishability and reduce bleeding. However, I've heard that fly ash retards setting, which would be undesirable when using the mix in cool weather. Is it possible to add fly ash without affecting the concrete setting time?*

A. Some fly ashes chemically retard setting. And many times fly ash is used in concrete as a cement replacement in combination with a water reducer. These concretes set more slowly than a straight cement mix would.

If you don't use fly ash to replace cement, but just to increase fines, it often won't retard setting. It serves as an additional fine aggregate and makes your patching mix easier to finish and less likely to bleed excessively. If the patching mix must be air entrained, you will probably need more air-entraining agent to get the same air content as a mix that doesn't contain fly ash.

Fly ash and drying shrinkage

Q. *We're the contractor on a large concrete repair project, and the mix proportions for the repair concrete include replacing 18% of the portland cement with fly ash. A co-worker told me that fly ash increases drying shrinkage and now I'm concerned about potential cracking. What effect does fly ash have on the drying shrinkage of concrete?*

A. According to the Portland Cement Association's Design and Control of Concrete Mixtures, the effect of fly ash on the drying shrinkage of concrete when used in low to moderate amounts is of little practical significance.

Perlite and its uses in concrete

Q. *We know that perlite is used in lightweight concrete, but where does perlite come from and how is it produced?*

A. Although it sounds like one, perlite is not a trade name. Rather, it is a generic term for naturally occurring siliceous volcanic glass.

The feature that sets perlite apart from other volcanic glasses is that when perlite is heated quickly above 1,600° F, it expands from 4 to 20 times its original volume. The expansion is caused by the presence of 2% to 6% combined water in the crude perlite rock. When heated, the crude perlite pops in a way similar to popcorn as the combined water vaporizes and creates numerous closed spheres in the heat-softened glassy particles. These air bubbles account for expanded perlite's light weight, which can be as little as 2 pounds per cubic foot. The maximum weight of expanded perlite is generally around 25 pounds per cubic foot. When used in concrete, perlite generally has a loose, dry bulk density of 7½ to 12 pounds per cubic foot.

Because of perlite's outstanding insulating characteristics and light weight, it is widely used as a loose-fill insulation in masonry construction. When perlite is used as an aggregate in concrete, a lightweight, fire-resistant, insulating concrete is produced that is ideal for roof decks and other applications. Perlite also can be used as an aggregate in portland cement and gypsum plasters for exterior applications and for the fire protection of beams and columns.

For more information, contact the Perlite Institute Inc., 88 New Dorp Plaza, Staten Island, NY 10306 (718-351-5723).

What is shrinkage compensation?

Q. *Manufacturers often claim that their repair mortars exhibit little or no shrinkage and supply test results to back up this claim. However, many of these mortars attain strengths of 8000 psi or greater, which leads me to believe that they contain a fair amount of cement. When asked how the mortars can have such low shrinkage values, manufacturers tell me that the mortar is "shrinkage-compensated." What does this mean?*

A. Shrinkage-compensating mortar typically contains a special cement or expansive agent that causes the mortar to increase in volume after setting and during hardening. This expansion offsets, or compensates for, the subsequent drying shrinkage. Ideally, the amount of expansion will equal the amount of drying shrinkage, resulting in an overall volume change near zero (see graph).

For shrinkage compensation to be effective, the mortar must be restrained by the patch edges or reinforcement, so that the hardened mortar goes into compression during expansion. Then the subsequent drying, instead of inducing tensile stresses in the mortar, which might result in cracking, merely reduces or relieves the compressive stresses caused by the initial expansion.

To determine whether low shrinkage values translate into improved performance in the field, Alexander Vaysburd of Structural Preservation Systems, Baltimore, is testing 12 repair

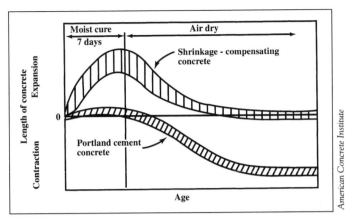

Typical length-change characteristics of shrinkage-compensating and normal portland cement concrete.

mortars from eight manufacturers and comparing laboratory results with field performance. The results of this research should shed some light on the effectiveness of shrinkage-compensated mortars.

Reader response: As a member of ACI 223, Shrinkage-Compensating Concrete, I led the drive to get the committee to elucidate the function of restraint in developing shrinkage compensation. This was over 20 years ago. As a result of the committee's deliberations, ACI Standard 223-77 was written, which clarifies the question of how much restraint is required. The answer is that as little as 0.15% steel will do the job, but more than 2% is probably too much. If the concrete or mortar cannot stretch the steel enough to equal the subsequent drying shrinkage, compensation is not provided. The restraint, in other words, must be resilient. Relatively immovable restraint, such as, say, 5% steel or the sides of a repair cavity, cannot be deformed enough by the early-age cementitious matrix to provide for the subsequent drying shrinkage. The important parameter here is concrete strain, not stress.

William F. Perenchio, Senior Consultant, Wiss, Janney, Elstner Associates Inc.

Q. *What is hot cement? When does it occur? Does it affect our concrete?*

Producers can adjust for hot cement

A. The American Concrete Institute (ACI) defines hot cement as newly manufactured product that has not had an opportunity to cool after burning or grinding. Freshly produced, cement is discharged from the kiln at temperatures near 180° F. In most situations, manufacturers store cement for several days before shipment, and its temperature approaches that of ambient air. Sometimes strong customer-demand reduces this cooling period, and cement has a higher temperature when it reaches the end user.

During the 1950s the Portland Cement Association conducted extensive laboratory and field tests on concrete produced with hot cements. On concrete pavement projects that used cement with temperatures as high as 167° F at the mixer, researchers found no significant effect on compressive or flexural strengths, volume change, checking or cracking when mixwater temperature was adjusted to compensate for the hot cement. The researchers concluded that the temperature of the freshly mixed concrete, and not the temperature of the separate ingredients, affects concrete properties (Ref.1).

However, hot cement can contribute to slump loss, especially if the cement has false- setting tendencies. Hot cement can also increase water demand and increase rate of slump loss by raising concrete even though it comprises only 10 to 15 percent of concrete's weight. Elevated cement temperatures will increase concrete mix temperatures about 1° F for each 8° F increase in cement temperature. Producers are advised to establish a maximum limit of 170° F on cement temperature, as it enters the concrete (Ref. 2).

To manage concrete temperatures, one concrete producer equipped drivers of his bulk-haulers with hand-held infrared heat meters. At the cement terminal drivers would first measure all the silos' temperatures, and then load material from the cooler ones.

References

1. William Lerch, "Hot Cement and Hot Weather Concrete Tests," Portland Cement Association, 1954.

2. ACI 305R-91, "Hot Weather Concreting," American Concrete Institute, Farmington Hills, Mich., 1994.

Using seawater in concrete

Q. *Our ready mix plant is located near the ocean and we were wondering if seawater can be used as mixing water in concrete. How would seawater affect the properties of concrete?*

A. It has been reported that using seawater can speed up the set of concrete and moderately reduce ultimate strength. Another potential problem is efflorescence, or the formation of salt deposits on the surface of the concrete. However, there are some accounts of unreinforced concrete made with seawater performing well for more than 25 years. Because of the danger of rebar corrosion, seawater should not be used in reinforced or prestressed concrete.

Strand chucks require inspections

Q. *The chuck bodies I inspect each day have been hit many times with a hammer. Does this lower the capacity of the chuck body?*

A. Most manufacturers of chucks test their bodies to at least 125% of the ultimate stand capacity. This gives more than adequate safety margin for the bodies. But each time a heavy steel hammer hits the chuck body, it leaves a dent in the body. After the body receives several blows, it may be weakened enough to cause the body to split and release the strand. This is especially true if the blows are on the narrow cap end of the body. Hammering on the bodies should not be necessary if the parts are cleaned and treated with a release agent recommended by the manufacturer. If you do hammer the body, use only plastic- or rawhide-tipped hammers.

Chapter 14

Mixing

Q. *When our ready mix trucks discharge concrete, drivers sometimes report seeing balls of nearly dry material coming down the chute. When broken open, the balls appear to be mainly cement and sand with some coarse aggregate particles in them. What causes this to happen and how do we prevent it?*

A. Richard Meininger of the National Ready Mixed Concrete Association says these balls are usually caused by a dry pack of cement and sand in the hub end of the mixer. The material stays there during the mixing cycle and doesn't break loose until the truck starts to discharge. The clumps get rounded into a spherical shape by the tumbling action in the drum as they move to the discharge end.

Because this is caused by getting too much fine material packed into the hub end of the mixing drum, the solution is to get some water and coarse aggregate into the hub end first. When charging the mixer, first batch part of the coarse aggregate and water. This will help prevent dry pack and should solve your problem.

Q. *We've noticed that concrete from onsite mixers often stiffens within minutes of pouring, then loosens up later. What causes this?*

A. You're seeing an example of false set—a pronounced loss of plasticity without much heat evolution shortly after mixing. It's mainly a nuisance rather than a major problem because plasticity can be restored by remixing or vibrating the concrete. The concrete will then set normally.

False setting is more likely when onsite mixers are used. That's because ready-mixed concrete is usually being mixed or agitated during the period when false setting occurs. But sometimes, when a set-retarding admixture is used with a cement that has false setting properties, false set occurs after ready-mixed concrete has been discharged.

Several possible mechanisms of false setting are described by Mindess and Young in their book *Concrete*, published by Prentice-Hall. One explanation is that false set is caused by dehydration of gypsum added to cement during manufacture to control setting of tricalcium aluminate. At high temperatures during the cement grinding process, water is driven from the gypsum, converting some of it to plaster (calcium sulfate hemihydrate). When water is added during mixing of concrete, the plaster rehydrates back to gypsum, forming a matrix that stiffens the mix. Because there's so little gypsum in the cement, the effect doesn't last long and further mixing destroys the matrix.

According to Mindess and Young, there is also evidence that set-retarding admixtures can accelerate the initial hydration of tricalcium aluminate and thus cause false set by increasing the formation of ettringite. In some high-alkali cements the formation of syngenite may cause false set. But it has also been found that false set sometimes occurs even when the crystallization of gypsum, ettringite, or syngenite can't be implicated.

Clumps of dry material in ready mix discharge

Onsite mixers and false setting

Maximum allowable time from initial batching to placement

Q. *I am involved with a large urban concrete project where concrete delivery is sometimes delayed by traffic congestion. The specification requires concrete placement within 90 minutes of batching at the plant. What is the significance of this 90-minute specification and should this be used as a rejection criterion?*

A. A concern regarding delivery and placement times in hot weather is the rate of slump loss in the concrete. One study showed that after 90 minutes of mixing, concrete at 90° F had a 2¾-inch slump loss (*ACI Journal*, August 1977). Specifications on delivery time usually aim to limit the amount of slump loss, so concrete can be placed and finished without adding excess water to the mix.

ASTM C 94, "Standard Specification for Ready Mixed Concrete," states that the time from concrete batching to placement should be 90 minutes or the time it takes for 300 drum revolutions, whichever comes first. However, it also states that this requirement may be waived by the purchaser if, after 90 minutes, the concrete still can be placed without adding water to the batch. The 90-minute window is not precise. The key is getting workable concrete without exceeding maximum water requirements. Even if the slump has fallen out of specification, ACI 305, "Hot Weather Concreting," allows for additional water to increase slump as long as the maximum allowable water content or water-cement ratio is not exceeded. High-range water-reducing admixtures also can be used to return the concrete to a workable condition if excessive slump loss occurs.

Can cement balls cause cracks?

Q. *I arrived at the job near the end of the day's slab placement to find the finishers picking clumps of cement out of the mix during finishing. They continued finishing, and the floor is now in place. I was upset with the foreman on the job for not rejecting the load. What caused the cement to ball up like that and will it have an effect on cracking in the future?*

A. The clumps of cement usually are referred to as "cement balls." Al Litvin from the Portland Cement Association says that the balls usually are caused by improper batching sequences and can lead to different types of cracking, depending on ball size and consistency. If the balls are large, the cement near the center may never contact water and remain unhydrated, thus reducing the strength in the balls considerably. Large cement balls also can reduce strength in the rest of the slab by consuming a majority of the cement. Cores should be taken to ensure that the concrete is of sufficient strength.

If the cement balls are small and completely hydrated, they can result in localized areas of paste that are prone to drying shrinkage cracks.

Chapter 15

Overlays

Q. *We're doing a rehab job on an old building that's being converted into offices. We tore out a 2-inch-thick wood floor that was too creaky. Beneath it we found 2x8-inch sleepers on 16-inch centers. The voids between sleepers were filled with what looks like a cellular concrete. The owner wants us to place a 2-inch slab of concrete over the existing sleepers and cellular concrete. We can't place a thicker slab without removing part of the sleepers and increasing the cost substantially. The slab will be covered with carpet but will occasionally have hard-wheeled cart traffic on it (maximum load of 2,200 pounds on four 6-inch-diameter wheels). The sleepers are in good condition. Should we place a polyethylene slip sheet down before placing the concrete? Or should we try to bond the 2-inch topping to the subfloor? Is there a chance that the topping will crack badly and make the floor unusable? In places it might even be a little less than 2 inches thick.*

A. We asked several engineers and contractors for ideas on this problem. A bonded topping would probably work better than an unbonded topping with a slip sheet because of the 2-inch thickness. Gene Boeke, an Atlanta contractor, suggests that you use what he calls the heel test on the cellular concrete to determine its bearing capacity. Put all your weight on one heel and note whether or not it makes an indentation in the material. If not, you probably have a bearing capacity of at least 2,000 pounds per square foot which isn't too bad. It should give the topping adequate support.

Boeke thinks that curling could be a problem with the thin topping. He suggests using low-slump topping concrete with a compressive strength no greater than 3000 psi to minimize shrinkage and curling. Cutting joints no more than 10 feet apart both ways will also help to control curling. Putting sheets of welded wire fabric at about midheight in the topping will keep random cracks between joints from getting wider. Don't run the wire fabric through the joints.

Another approach is to place the topping on steel decking with a ½-inch rib. This gives the topping added tensile strength and more uniform support. Minor cracking isn't a problem from an appearance standpoint because there's carpeting on the floor. But faulting at the cracks under the heavy cart load might cause a problem. The decking would eliminate that problem, but would cost more than just placing the topping on the fill. Boeke also warns that the cellular concrete might contain slag that could cause the steel decking to corrode.

Q. *We need to place an overlay on existing concrete to correct a grade problem. The existing concrete was placed last week and still has a pigmented curing compound on the surface. An acid etch didn't remove it. Does it have to come off before we place an epoxy bonding agent? The area is about 3 feet wide by 400 feet long.*

A. Remove the curing compound. Otherwise the overlay may not bond properly. Try using mechanical methods such as shotblasting or scarifying.

Concrete topping for cellular concrete

Remove curing compound before placing an overlay

Coating a floor that contains fiber reinforcement

Q. *We plan to apply a coating to an industrial floor that contains synthetic fiber reinforcement. We acid etched the floor, but after the treatment was completed, fibers were sticking out of the surface. Before we apply the coating, how can we remove the fibers so they do not protrude from the floor?*

A. When fibers are used in concrete, they should be equally distributed throughout the concrete matrix or paste. Since finishing brings paste to the surface, it should be expected that fibers will be embedded close to the surface. Before coatings are applied, floor surfaces are usually shotblasted and sometimes acid etched to improve coating bond. Acid etching, however, can expose some of the fibers. The Synthetic Fiber Association and many fiber manufacturers recommend shotblasting rather than acid etching because of the safety hazards associated with using acid.

Fiber manufacturers say protruding fibers can improve coating bond through the mechanical interlocking of the coating to the concrete. However, the fibers may be visible through some coatings. If this is objectionable, you can remove the fibers by burning them with a torch before applying the coating. If the use of an open flame is not allowed, you can also remove the fibers with a high-pressure water stream or by shotblasting. Initial surface preparation by shotblasting instead of acid etching can save you from having to do this additional work.

Why use whitewash before whitetopping?

Q. *I've seen workers whitewashing an asphalt pavement before placing a concrete overlay. What's the purpose?*

A. The whitewash serves two purposes. It cools the surface during hot weather so concrete in contact with the asphalt doesn't set too fast. And it serves as a bond-breaker between the pavement and the overlay.

Should whitetopping have uniform thickness?

Q. *Your August 1990 article on whitetopping a driveway says that compacted fill was placed in low areas because shrinkage cracks might start at points where the thickness changed. I attended a demonstration by the Michigan Concrete Association where the same practice was shown. A controlled density fill was placed in low areas.*

Then Robert Huber reported on 9 miles of whitetopped pavement in Iowa where all the ruts and chuckholes were filled with concrete as the whitetopping proceeded. He says the extra thickness of concrete at these points is beneficial—and his pavement has been in service almost 13 years. Who's right?

A. We asked Dale Diulus of Phoenix Cement to comment because he has been involved in many whitetopping projects. He gives two reasons for not filling chuckholes at the same time the overlay is placed. The variable cross section restrains overlay movement and causes cracking. Also, reentrant corners in the pavement cross section may induce cracking.

That said, however, we can't argue with success. The pavement in Iowa has performed well. So perhaps both methods can work.

Thin, unbonded topping over plywood?

Q. *The plywood diaphragms under the floor of an unreinforced masonry building are warped, producing an uneven floor surface. I want to place a ¾-inch-thick concrete topping to improve floor flatness, but a bonded topping would stiffen the diaphragms and change the load paths. Can I use a bond-breaker on the plywood so the topping will be unbonded?*

A. An unbonded topping only ¾ inch thick is likely to crack extensively. But if it will be covered with carpeting or another floor covering that won't reflect the crack, this may not be a problem. Ask the manufacturers of cement- or gypsum-based floor underlayments for their recommendations, based upon the floor covering you plan to use.

Q. *With more and more stringent environmental regulations, is acid etching still a viable way to prepare concrete slabs before applying a coating?*

A. According to Robert Cain of KRC Associates, acid etching is an effective and safe procedure. You must first remove curing compounds, sealers, and any oil or grease. Then spray a 10% solution of muriatic acid uniformly over the surface. Foaming and bubbling occur and the acid is neutralized as it attacks the surface. Flush the surface with water. Because the acid is neutralized by alkalies in the concrete, the flushing water can be safely washed down a floor drain. After etching, the floor surface should feel like coarse to medium sandpaper.

Q. *We are involved in a warehouse renovation project. The new warehouse tenants will be moving heavy machinery and forklifts into the facility, and the structural engineer has determined that the 6-inch-thick concrete floor is not thick enough to support the additional weight. Otherwise, the floor is in good condition. The structural engineer recommends placement of a 2-inch-thick bonded overlay on the floor. Do you have any recommendations on mix proportions for the overlay, especially aggregate sizes? Should the overlay be reinforced? What about joint spacing?*

A. We spoke to Jerry Holland of Lockwood-Greene Engineers and he had several recommendations. As far as mix proportions, it's very important to keep the water content of the overlay as low as possible to minimize shrinkage. Shrinkage of cementitious toppings is a major concern for two reasons. First, thin concrete sections tend to shrink more than thicker sections. Second, the topping will be susceptible to curling because the underlying concrete will not absorb as much water from the overlay as an aggregate base. As a result, the overlay surface will dry faster than the bottom portion of the overlay. As the surface dries, it shrinks, and the overlay curls upward at the edges. This may cause debonding of the overlay. Holland stressed that the mix should contain uniformly graded aggregates rather than gap-graded aggregates. He said that he has had success with similar overlays using a combination of #67 and #89 aggregates (ASTM D 448).

Because welded wire fabric would be difficult to handle and position in a 2-inch-thick overlay, use fiber reinforcement. Holland recommended using steel fibers instead of synthetic fibers for these service conditions because they provide more structural reinforcement and would help hold the concrete together in areas where the overlay does not achieve a good bond to the substrate.

The overlay joints must be sawed directly over the underlying floor joints, and Holland recommends sawing them full depth. According to Holland, an overlay joint and an underlying joint may begin and end at the same place, but they often are not aligned perfectly along the entire length of the joint. Sawing the overlay joint to its full depth reduces the chances of reflective cracking in the overlay in areas where the joints are not perfectly aligned.

Q. *We're installing a latex-modified concrete overlay on a deteriorated bridge deck. However, the existing deck has several patches of magnesium-phosphate-cement mortar. Will we encounter any bond problems in these areas or any other problems with compatibility of these two materials?*

A. We spoke to Bob Gulyas of Master Builders Inc. and he said that you should not experience any problems. Before placing the overlay, prepare the surface of the patches in the same manner as the rest of the deck.

UV resistance of epoxy overlays

Q. *An article on page 183 in Concrete Repair Digest, June/July 1995, described how epoxies can be used as thin overlays on bridge decks. I understand that epoxies are not resistant to ultraviolet (UV) radiation. If this is true, won't they degrade when exposed to sunlight on bridge decks?*

A. We spoke to Michael Sprinkel of the Virginia Transportation Research Council, and he said that very little of the epoxy used in bridge overlays is exposed to sunlight because most of the resin is shielded by the gap-graded aggregate that is broadcasted over it. Oil, dirt, and other debris provide additional protection for the epoxy. Sprinkel concedes that, if applied neat, the epoxy binders degrade quickly when exposed to UV radiation. However, epoxy overlay samples taken from several bridge decks show very few signs of UV degradation.

How to remove curing compound before applying topping

Q. *We're the general contractor for a repair project that includes applying an epoxy terrazzo floor coating on an existing concrete floor. The floor is 1 year old and was treated with a cure-and-seal product after it was poured. We're looking for an economical method to remove the cure-and-seal. The coating applicator says he wants the floor shotblasted and won't guarantee his work unless that's the surface preparation method used. However, we think chemical cleaning of the floor would be less costly. What chemicals can be used to remove cure-and-seal products?*

A. You need to find out what cure-and-seal product was used; then contact the manufacturer and ask what solvents will remove it. It's likely that toluene, xylene, or similar solvents will do the job.

In comparing the economics of different surface preparation methods, however, you may not be considering all the factors. Most solvents used to remove coatings are hazardous materials and require stringent safety precautions. You'll also have to arrange for disposal of the solvent. And if you're working with an inexperienced crew, the solvent cleaning may not be done thoroughly enough. How will you know if all the floor area has been treated? That's probably why the applicator won't guarantee his work unless the floor is prepared by mechanical methods.

Any savings resulting from using the chemical cleaning are likely to disappear if there's a bond failure of the epoxy product.

Thin overlay for parking structure ramp

Q. *On a parking structure project in the South a 3,000-square-foot area of ramp slab has a wavy surface. The depressions seem to be about ¼-inch thick. Can a cementitious overlay correct this?*

A. Be cautious about that approach. Our reservations about thin cementitious overlays in exposed traffic applications relate mainly to two factors: successful bonding and the potential for shrinkage cracking.

Bond is important for most cementitious overlays and critical for thin ones. If a bond failure occurs, the delaminated overlay may crack and spall under traffic. To develop good bond, extensive surface preparation—deep shotblasting, sandblasting, or scarifying—may be needed.

The surface usually is conditioned with a bonding agent, grout, or paste fraction of the patch material just before applying the repair material.

Careful curing is vital to avoid shrinkage cracking. A prolonged moist cure is recommended for cementitious mixes; latex-modified repair materials should be cured in strict accordance with manufacturer recommendations.

Before proceeding much further, get a better idea of the depth and extent of waviness. Rather than adding material, you may have more success grinding the high spots.

Q. *We recently placed a 3-inch-thick overlay for a small heated indoor parking area. We first routed existing cracks and filled them with an epoxy material. Next, we shotblasted the existing concrete surface, then broomed in a proprietary bonding agent. Then we placed 5000-psi pea gravel having a 4- to 5-inch slump. The owner observed our work very carefully and was pleased that no cracks had appeared a month after we did the work.*

We were called back a month after the work was done, however, because the edges of the overlay had debonded. Also, cracks had formed perpendicular to the edges over about half of the overlay perimeter. The cracks are about 2 feet long and are spaced 4 to 8 feet apart.

We plan to remove and replace a 2-foot-wide section of the overlay around the perimeter, but we don't want to make the same mistake twice. What could have caused the loss of bond and cracking?

Causes for overlay debonding

A. There are several possible causes for the debonding. If too much time passes between applying the bonding agent and placing the overlay, the bonding agent could act more like a bond breaker. Another possible cause is excessive drying shrinkage of the overlay concrete. When the heat came on in the garage, drying and shrinkage of the concrete was accelerated. Pea gravel concrete is more shrinkage-prone because it requires larger amounts of mixing water than concrete with ¾-inch aggregate or larger. More water means more drying shrinkage. The shrinkage causes curling that can "unzip" the overlay at the edges. It also causes cracking.

You didn't mention curing, but that's a critical factor affecting overlay performance. According to Bill Perenchio of Wiss, Janney, Elstner in Northbrook, Ill., concrete bond strength doesn't develop as fast as compressive strength. If curing is ineffective or is discontinued too soon, loss of bond is likely. Perenchio says the only sure way to cure overlays is with water applied externally, followed by wet burlap covered with plastic sheeting. He says a curing compound isn't enough to ensure proper curing.

We wouldn't recommend removing and repairing concrete at the overlay edges until you see whether the rest of the concrete stays bonded. Use a chain drag or soundings with a hammer to find out how far back from the edge the concrete has debonded. Then check it again in a month or two. If debonding is progressive, you may need to replace all of the overlay.

Reader response: I recently repaired a debonded, 3-inch-thick overlay that also was severely cracked due to high shrinkage. The debonded sections were identified with a chain drag. One of the most successful repair methods we used was to drill small holes into the debonded sections and pump a low-viscosity epoxy between the overlay and the underlying slab.

This worked very well and resulted in a completely bonded overlay. The overlay was at least 1 year old, so almost all of the shrinkage had taken place. The cracks were filled with sand and epoxy afterwards.

The above method was not cheap, but it was less expensive than removing the overlay, cleaning the floor, and recasting the concrete. Also, reuse of the slab was a matter of a couple days, not 28 days.

C. Kuilman, P.E., K. C. Design & Construction Management

Resisting concentrated nitric acid

Q. *Our firm recently inspected the floor of a potassium nitrate fertilizer plant. Four large process vessels have leaked hot, concentrated nitric acid onto the floors. Several areas of the floor have deteriorated and, after repairing the floor, we want to install a protective overlay. We considered installing acid brick with furan mortar, but the floor cannot accommodate the 2½-inch rise in elevation. Is there another overlay system we can use?*

A. Another overlay system that is resistant to strong organic acids is potassium silicate. Potassium silicates can be installed about 1½ inches thick and can perform in service temperatures as high as 2,000° F. Because potassium silicates are somewhat brittle, consider installing the system over an asphaltic membrane to protect the concrete should the overlay crack.

Coating a floor containing synthetic fibers

Q. *We've been hired to apply an epoxy coating to a warehouse floor that contains polypropylene fibers. The fibers stick up out of the concrete, especially after it's shotblasted. We don't want to coat over the fibers because we're afraid that we'll get voids in the coating. Can the fibers be removed with a floor sander?*

A. Bob Cain of KRC Associates says that polypropylene fibers are too soft and flexible to be removed by sanding alone. However, if you apply an epoxy primer to the floor, the fibers become rigid and stick up out of the floor after the primer cures. They can then be removed by sanding.

Another method is to burn off the fibers with an acetylene flame blaster. Be careful because flame blasters can fracture the surface of the concrete. Cain says that by using a low-intensity flame and moving quickly along the floor, you can burn off the fibers without damaging the concrete. Remember that flame blasting presents an obvious fire hazard and may generate toxic smoke and fumes depending on the contaminants on the concrete surface.

Preventing cracks in overlays

Q. *What is an effective means for preventing reflective cracks in thin unbonded overlays?*

A. The American Concrete Pavement Association indicates that bituminous-based materials provide the best results as an interlayer material. The mix should meet standard design specifications and should employ conventionally graded aggregates. The maximum aggregate size should be about half the minimum thickness of the separation course. ACPA suggests 1 inch as a generally adequate thickness for a bituminous separation layer.

Coating a metallic dry-shake hardened floor surface

Q. *We're a bleach manufacturer, and we've just acquired a facility with a floor that contains a metallic-aggregate surface hardener. We want to install a high-build epoxy coating on the floor to resist bleach spills. How should we prepare the floor for coating?*

A. Because metallic-aggregate floors are so resistant to impact, it's difficult to mechanically roughen the surface to prepare it for coating.

But Bob Cain of KRC Associates Inc. said that aggressive shotblasting, while not removing the metallic-aggregate cap, should roughen the floor enough to provide an adequate bonding surface. To achieve a rougher surface texture, Bob Gulyas of Master Builders Inc. recommended loading the shotblaster with steel grit instead of steel shot.

Another approach was suggested by Jay Cullinan of Floortech. He says he has had success preparing metallic-aggregate floors with 20% phosphoric acid. Don't use hydrochloric (muriatic) acid because it will corrode the metallic surface.

Before using any method on a large scale, it's always wise to prepare small sections of the floor, apply the coating, and perform adhesion tests as described in Appendix A of ACI 503R-93, "Use of Epoxy Compounds with Concrete."

Chapter 16

Paving

Q. *Is a granular base needed beneath city streets or parking lots that will be paved with concrete? Some specifications require a subbase and some don't. If one is needed, how thick should it be?*

A. There's an excellent article on this subject in the August 1989 *Ohio Paver*. Concrete pavements built on properly compacted and shaped subgrades, with no subbase, provide satisfactory performance for many years with little maintenance. Using a subbase adds significantly to the cost of a concrete pavement but doesn't appreciably increase its load-carrying capacity. Thus, a subbase may not be cost-effective and shouldn't be specified and used unless warranted for other reasons.

There are four reasons for using a subbase under concrete pavements:

- To prevent mud pumping (ejection of fine material from under the pavement) at joints and cracks
- To help control volume changes in highly expansive subgrade soils
- To help reduce excessive differential frost heave
- To provide a working platform for pavement construction

The primary purpose of a subbase is preventing mud pumping. Mud pumping occurs at joints when the subgrade soil goes into suspension, when there's free water between the pavement and subgrade, and when there's heavy truck traffic. The *Ohio Paver* article defines heavy truck traffic as 100 to 200 trucks per day, excluding panel, pickup, and four-tire single unit trucks.

If any of the three conditions that lead to mud pumping is absent, pumping won't occur and a subbase isn't warranted. The other three reasons for subbases are secondary. Volume changes and frost heave can usually be controlled by proper earthwork and compaction methods. These methods also provide the needed working platform for paving.

Regardless of whether or not a subbase is constructed, shaping and compacting the subgrade is important. Porous subbases permit water to enter between the pavement and subgrade. The water must flow out of the pavement structure, transversely or longitudinally. Low spots will collect water that weakens the subgrade.

The article recommends specifying subbases under pavements only when they're needed to assure satisfactory performance. Adequately designed concrete pavements without subbases are suitable for most city streets, county roads, low-traffic-volume highways, and general aviation airport runways. Subbases aren't warranted for most concrete parking areas, sidewalks, bike paths, golf cart paths, driveways, and patios.

Q. *We do everything possible to prevent cracking of the concrete driveways and patios that we place. We promptly cut joints to the correct depth and space them no more than 12 feet apart. We also use medium-slump concrete and start curing it as soon as we can without marring the surface. However, we still get an occasional crack, and the homeowner often wants us to repair the crack or remove and replace the cracked panel. Is there any information we can give homeowners to reassure them that a crack won't affect the structural function of their driveway or patio?*

Subbases not needed beneath most concrete streets and low-volume highways

Cracks in driveways: what's acceptable?

A. Let's start with the most common remedial requests from homeowners. Trying to repair a driveway crack by injection or patching usually makes the crack look worse than it did before the repair. And if you replace a panel, you'll almost never match the color of the surrounding concrete. So both of those fixes are likely to be objectionable from a cosmetic standpoint.

The effect of cracking on the structural performance of a driveway depends on crack width. Load transfer by aggregate interlock becomes ineffective when crack widths exceed about 0.04 inch. Thus, if the crack width is less than 0.04 inch (about the diameter of a standard paper-clip wire) the driveway can still provide adequate load transfer across the crack. An easily used clear-plastic crack-width gauge enables you to measure width at several locations along the crack so you can calculate an average crack width. You can get this free gauge by calling Construction Technology Laboratories Inc., Skokie, Ill., at 847-965-7500 and leaving your name and address.

We suggest showing owners photos of driveways with repaired cracks or replaced panels to illustrate the cosmetic problems associated with both of these approaches. Then, if you can demonstrate that the crack isn't wide enough to affect driveway performance, you can make a strong case for the use-as-is solution.

Pavement acceptance specification based on core properties

Q. *We're bidding on a local concrete paving job for which acceptance and payment are based on concrete compressive strength and pavement thickness, both determined by drilling and testing cores. At least two cores are taken each day. Design thickness is 6 inches, with a minus tolerance of 0.2 inch. Design compressive strength is 4000 psi, with a minus tolerance of 200 psi.*

If any core thickness or compressive strength measurement is less than the permitted value, the owner has two options:

- *Require us to remove and replace the portion of the pavement represented by the core*
- *Leave the pavement in place but deduct a percentage of the bid price as shown in the table on this page.*

According to the specification, slabs found to be deficient by core testing will not be recored for verification. American Concrete Institute documents consider concrete in the area represented by a core test to be adequate if the average core strength equals at least 85% of the specified strength. This specification specifically states that the 85% value is not acceptable.

This is the first time we've seen this kind of specification. Is there a way to bid the job without a high probability of losing money?

A. Quality control of the concrete production and paving operations must be outstanding if you bid this job. You'll also have to plan on a higher-than-normal overdesign for concrete strength.

Let's look at strength requirement first, using a modification of statistical methods described in Reference 1. Assume that your concrete supplier can produce concrete of the required strength with a standard deviation of 400 psi, based on at least 30 tests. This standard deviation indicates very good control. The required average strength can then be computed as follows:

$$f_{cr} = f_m + tS$$

where

f_{cr} = required average strength (psi)

f_m = specified minimum strength (psi)

t = a constant that depends on the number of tests that may fall below f_m

S = forecast value of the standard deviation (psi)

If you're willing to accept one chance in 100 of not getting paid the full bid price, $t = 2.33$ and specified minimum strength is $4000 - 200 = 3800$.

Thus: $f_{cr} = 3800 + 2.33 (400) = 4700$ psi

This compressive strength doesn't sound unreasonable. However, the producer's strength and standard deviation data are based on tests of standard-cured molded cylinders, not drilled cores. Research has shown that 28-day-old cores from a well-cured slab, tested dry, can be expected to attain about 85% of the strength of 28-day-old standard-cured molded cylinders

Thickness Deficiency (in.)	Deduction (percent of bid price)
Up to 0.20	0
0.20 to 0.40	15
0.40 to 0.60	60
More than 0.60	100

Strength Deficiency (psi)	Deduction (percent of bid price)
Up to 200	0
201 to 333	20
334 to 669	40
More than 669	100

(Ref. 2). If core strength is assumed to be 85% of the cylinder strength, a cylinder strength of about 5500 psi will be needed to produce a core strength of about 4700 psi. And even with the 38% overdesign on compressive strength, you'll still have one chance in 100 of a failing core-strength result.

Also, be aware that basing strength acceptance on core-test results is almost sure to cause problems if the results are low and you don't produce your own concrete. The concrete producer can justifiably argue that he isn't responsible for core-test results because core strengths are affected by factors over which he has no control. These factors include field consolidation and curing of the concrete and pavement coring methods. Be prepared for a dispute if there are low core-test results.

You can use a similar statistical approach to determine the required average slab thickness, assuming that core lengths have a normal distribution. Slipform paving contractors with experienced crews and rigorous quality-control programs have consistently obtained standard deviations on pavement thickness of 0.4 inches. Again, if you're willing to accept one chance in 100 of not getting paid the full bid price, t = 2.33 and specified minimum thickness is 6 − 0.2 = 5.8 inches. Required average thickness is then:

5.8 + 2.33(0.4) = 6.7 inches

To summarize, your bid price would be based on:
- Ordering 5500-psi concrete from a concrete supplier with very good production control. You must also consolidate and cure the concrete well.
- Attaining an average pavement thickness of at least 6.7 inches with closely controlled fine grading, a stable slipform trackline, carefully set stringline, uniform-slump concrete, good equipment, and an experienced paving crew. Any deviations from the needed quality control will increase your thickness standard deviation and require a larger average pavement thickness.

What's the probability of losing money? You'll have to answer that question.

References

1. ACI Committee 214 "Recommended Practice for Evaluation of Strength Test Results of Concrete" (ACI 214-77, Reapproved 1989), American Concrete Institute, Farmington Hills, Mich., 1989.

2. Delmar Bloem, "Concrete Strength in Structures," *ACI Journal*, March 1968, pp. 176-187.

Q. *Has anyone ever compiled a list of things a contractor can do to prevent problems that come up unexpectedly during a concrete pour? For example, our ready mix company has had trucks unable to get close enough to discharge the load, we have seen forms deflect and require extra concrete, and we*

Preparing for a concrete pour

have had trucks arrive on the site only to discover that the reinforcement still needed to be set in place. What are some things a contractor or homeowner can do to prepare for a pour? We would like to make this list available to our ready mix customers.

A. We suggest that you start by putting together a checklist of things you already know that a contractor or homeowner can do to prepare for a concrete pour. You can add to the list by talking to other ready mix suppliers about their experiences. Here is a partial collection to get you started:

- For slabs on grade, be sure the side forms are in place, secure, and checked for grade. Compact the subgrade and measure the depth of the slab from the top of the form to the subgrade surface by extending a straightedge from one side form to the other. Calculate the amount of concrete needed and then allow a slight amount of extra concrete to avoid having to order more.
- Be sure that reinforcement is in place and clean.
- Have the expansion joint material available on the site.
- In hot, dry weather, moisten the subgrade and forms.
- Be sure that the screed, straightedge, and finishing tools are out and accessible.
- Even if it doesn't look like rain, have plenty of plastic sheeting on hand to cover the pour.
- Provide a clear path for the ready mix truck, free of obstacles and debris, and be sure that it is capable of supporting the heavy load of the truck and the concrete.
- If trucks will be following one another, provide room for a truck to back out and leave while one or more others are waiting.
- If accessibility to forms is limited, confer with the ready mix supplier about providing chute extensions or using a conveyor or pump to get the concrete to the forms.
- Make a final check to see that the forms are well braced so they will stay in alignment, won't require extra concrete, and will be safe.
- Be sure the curing material is on hand and in an adequate amount.

How deep does microcracking penetrate?

Q. *We're involved in a bridge overlay project in which the concrete cover over the reinforcing steel will be milled before placing the overlay. Before full-scale construction, a small section of the bridge deck was milled and an overlay was placed as a test area to evaluate the in-place, direct tensile bond of the repair. Many of the cores failed below the minimum required tensile-bond strength of 200 psi. Of those cores, most of them failed within the substrate concrete, and we suspect that microcracking, or bruising, of the substrate during milling may have been a contributing factor. How deep into the substrate must failure occur for microcracking to be written off as a cause of failure?*

A. We spoke to Michael Sprinkel of the Virginia Transportation Research Council, and he had several observations concerning tensile-bond testing and modes of failure for concrete overlays. Sprinkel said that, in his experience, tensile-bond testing of overlays placed on properly prepared substrates usually results in a break in the substrate concrete near the bottom of the drilled core. If specimens consistently fail at this depth at strengths lower than the minimum specified, this usually indicates weak substrate concrete. In this case, you should consider replacing the deck because placement of an overlay will most likely result in delamination.

If, however, specimens fail in the substrate concrete within ¼ inch of the bond line, microcracking is often the cause. Sprinkel said that microcracking often can be remedied by shotblasting or grit blasting this weakened layer off the concrete surface. Breaks at the bond line are usually caused by oil or other bond-inhibiting substances that have not been adequately removed from the substrate.

Chapter 17

Placing

Q. *We're placing concrete for a job with 10-foot-high columns, 18x18 inches in cross section. The columns contain four #8 vertical bars. The fastest way to place the column concrete is by direct discharge because there's an unobstructed path to the bottom of the column. After we'd placed several columns this way, the engineer said we had to use a tremie hose because concrete shouldn't free fall more than 3 to 5 feet.*

ACI 301 is the specification for this job and I can't find any mention in the specification of a height limit for concrete free fall. We've already stripped columns and there's no evidence of segregation or honeycombing. Is it necessary to use a tremie under these conditions? And if so, what's the maximum permitted height for dropping concrete without a tremie?

A. Some publications used to give a recommended free fall for concrete. However, ACI 301 makes no mention of a 3- to 5-foot limit.

ACI's "Guide for Measuring, Mixing, Transporting, and Placing Concrete" (ACI 304R-89) in Section 5.4.1 gives some precautions for placing. The guide says equipment should be arranged so that the concrete has an unrestricted vertical drop to the point of the placement. The stream of the concrete shouldn't be separated by permitting it to fall freely over rods, spacers, reinforcement, or other embedded materials. If forms are sufficiently open and clear so that the concrete isn't disturbed in a vertical fall into place, direct discharge without the use of hoppers, trunks, or chutes is usually desirable.

PCA's *Design and Control of Concrete Mixtures* (13th edition, page 104) specifically addresses your question. This source says the height of free fall of concrete need not be limited unless a separation of coarse particles occurs (resulting in honeycomb). If honeycomb does occur, a limit of 3 to 4 feet may be adequate.

The PCA book cites a study in which concrete was dropped vertically 50 feet into a caisson. There was no significant difference in aggregate gradation between control samples as delivered and free-fall samples taken from the bottom of the caisson. This showed there was no segregation.

Q. *I've heard of yield loss in high walls caused by loss of air due to vibration and compression of air because of concrete pressure before it sets. Could this explain an underyield of 10%?*

A. The maximum loss in yield if all the air is removed by vibration is limited to the initial air content. For non-air-entrained concrete, that's usually 1% to 2%—far less than the 10% loss you experienced. Even for air-entrained concrete, it's highly unlikely that you'd lose enough air or compress the air enough to cause a 10% yield loss. The initial air content would have to be more than 10%.

Sometimes there's an apparent yield loss because of form movement, bulging, or setting the forms incorrectly. However, you'd have to have a pretty big bulge to get a 10% loss. If a wall were

20 feet high, 150 feet long, and 10 inches thick it would require about 93 cubic yards of concrete. To cause an apparent yield loss of 10%, the average wall thickness would have to increase to 11 inches. If it did, 93 yards of concrete would fall about 9 yards short of filling the forms.

Will calcium chloride thaw a frozen subgrade?

Q. *We're supplying ready-mixed concrete for a floor slab that will be placed in freezing weather. We warned the contractor about placing concrete on a frozen subgrade. He says he'll throw flake calcium chloride on the subgrade surface to thaw it before he places the concrete. Is there any code or document that prohibits this?*

A. We don't know of a document that specifically prohibits this practice, but it's not an effective way to thaw the subgrade to any depth. Some contractors put up a tent or other enclosure and use space heaters inside it to thaw the subgrade (see *Concrete Construction*, November 1989, page 956). Others use electrically heated concrete curing blankets that are laid on the ground. One such product comes in 36-foot-long sections that are 4½ feet wide. The blankets require a 220-volt power source and can be covered with insulated curing blankets to retain the heat. They reportedly maintain a 60° F temperature beneath the blanket even when the air temperature is at 0° F. The cost is estimated at less than $2.50 per square foot.

Long-strip construction preferred for slabs on ground

Q. *We're building a slab on grade for a large warehouse. The engineers want us to place 5,000 square feet maximum for each pour and use a checkerboard sequence. We're able to place 10,000 to 15,000 square feet daily if we're allowed to use the long-strip method. Is there any literature that we can use to convince the engineer that our method will produce as good a floor?*

A. There's a widely held belief that checkerboarding eliminates or minimizes shrinkage cracking because earlier placements shrink before infill panels are placed. This supposedly reduces shrinkage stresses, shrinkage cracking, and joint widths. But according to the Portland Cement Association, the reasoning behind this idea is incorrect. Shrinkage of earlier placements occurs too slowly for the method to effectively reduce shrinkage and joint widths. Because drying shrinkage takes place over a long period, lengthy delays would be required between casting of adjacent bays to gain much benefit from checkerboarding. The American Concrete Institute's "Guide for Concrete Floor and Slab Construction" (ACI 302.1R-89) recommends using long-strip construction. Checkerboard construction is not recommended.

Time limit between pours to avoid cold joints

Q. *Is there an allowable time limit for successive placements of concrete during a monolithic pour? In other words, once one truckload has been discharged, how long can I wait before placing the next truckload?*

A. We don't find any reference to a set time limit in the American Concrete Institute (ACI) building code or specifications for structural concrete. That's probably because time isn't the only factor to consider. In cool, damp weather you can wait longer to place the next batch than you can during hot, dry weather. As long as the previously placed concrete is still plastic (an internal vibrator will penetrate it), you can safely place the next layer of concrete without producing a cold joint.

ACI's "Guide for Consolidation of Concrete" (ACI 309R-87) gives the following advice: When the placement consists of several layers, concrete delivery should be scheduled so that each layer is placed while the preceding one is still plastic to avoid cold joints. If the underlying layer has stiffened just beyond the point where it can be penetrated by the vibrator, bond can still be obtained by thoroughly and systematically vibrating the new concrete into contact with the previously placed concrete; however, an unavoidable joint line will show on the surface when the form is removed.

Q. *Two similar questions have been raised regarding placement of concrete on top of recently completed slabs and footings. In one case the contractor wanted to form and place his basement walls the day after completing the footings, but was being required by the architect to wait seven days. In the other case, the engineer was denying permission to place column concrete above a fully shored and supported slab that had been placed 5 to 7 hours earlier the same day. The columns in question were centered above those on the lower level. What are the rules that limit the timing of concrete pours under these conditions?*

A. After much searching and inquiry, we've concluded that there aren't any written rules to cover these cases. The concrete pretty much makes its own rules in terms of setting and hardening time, and builders respecting these natural limits have not had problems. For example, the contractor will not go on a finished slab to set formwork for columns until concrete is hard enough to remain undamaged by the activity. Formwork and shoring supporting the slab are normally designed also to carry loads from construction work on the slab.

Basement walls: It is common practice to pour walls the day after the footings are poured but you're not likely to find a reference that says either you can or can't do this. One reason this is comparatively safe is that footings are frequently sized to meet local code minimums that are actually oversize when compared with the load that the wall will place on the footing. Remember also that at this stage of construction the only load on the footing will be from the weight of the wall since the structure above for which the footing has been designed is not yet in place.

Placing columns on the new slab: After checking more than a dozen books, technical reports, and standards dealing with formwork and concrete construction (from both the United States and Europe), we found no statement limiting time of placement of columns on top of a slab.

What we did find, however, are restrictions on the related condition of placement of concrete in slabs and beams on top of a deep lift of fresh concrete in walls or columns. The Amer-

© R.W. Steiger

How soon after a slab is finished can work on columns begin? On some fast-cycle high-rise work builders go in later the same day to set rebar and forms. Concrete is often placed the same day or early the following day.

ican Concrete Institute (ACI) *Manual of Concrete Inspection*, the ACI "Specifications for Structural Concrete for Buildings (ACI 301-84)," and the ACI Building Code (ACI 318-89) all have statements similar in intent to the following Section 8.3.2 from ACI 301-84:

"Placing of concrete in supported elements shall not be started until the concrete previously placed in columns and walls is no longer plastic and has been in place at least two hours."

Not one of the three ACI documents sets any limit for the related condition of placement of concrete above the supported element (beam or slab).

We discussed the question with the former chairman of ACI Committee 301, David Gustafson, technical director of the Concrete Reinforcing Steel Institute; and with the current 301 chairman, Timothy Moore of Gilbert/Commonwealth Inc. Both said they knew of no rules or limitations on time of placement of column concrete on top of recently finished slabs. Moore further stated that no such provisions are being considered among the many changes the committee is considering for future revisions of ACI 301. He said that the only other provisions of ACI 301 which might have any bearing would be those for construction joints. Section 6.1 of ACI 301 (construction joints) has provisions for location of construction joints and bonding at construction joints where required or permitted, but nothing on timing of placement.

We also queried Randy Bordner, former chairman of ACI Committee 347, Formwork for Concrete. Bordner is a professional engineer and a specialist in form design and construction for multistory buildings. He stated that on his jobs column concrete has routinely been placed above slab concrete placed on the same day, the only concern being satisfactory hardness of the slab for attachment of any necessary templates and bracing.

Of related interest, P. Kumar Mehta of the University of California, Berkeley, says in his book *Concrete Structure, Properties, and Materials* regarding the setting and hardening of cement paste in concrete (page 191): "The time taken to solidify completely marks the final set, which should not be too long in order to resume construction activity within a reasonable time after placement of concrete."

This statement implies that resumption of construction activity could take place at the time of final set.

Temperature protection for concrete piles

Q. *I need to place concrete in four 30-inch-diameter, 37-foot-long steel-shell piles that are driven at the bent lines of a river bridge. River depth varies, but is 5 feet in the deepest spot. After the steel shell is driven, the water level is about 1 foot below the top of the pile. The water moves swiftly and its temperature ranges from 35° to 44° F.*

Specifications require me to keep the concrete temperature above 45° F for 72 hours and above 40° F for the next 96 hours. The concrete is a seven-sack, 3500-psi mix with no changes allowed in the mix proportions. Each pile will only take about 7 cubic yards of concrete, so I don't think heat of hydration will be sufficient to maintain the minimum temperatures. How can I handle this problem? The job schedule won't permit us to wait until the weather is warmer.

A. Since the embedded part of the pile will probably retain enough heat to meet the minimums, your main concern is the 4-foot section that's exposed to the cold water. Perhaps you could drive a slightly larger-diameter (and shorter) steel-shell pile around each pile to form a cofferdam. Then you could pump out the water and wrap the inner pile with a closed-cell insulating foam. Another possibility: Leave the water in the annular space between the two sheet piles and heat it by using an electric immersion heater or, more safely, by circulating hot liquid through tubing placed in the annular space.

Note: The specifying agency wouldn't permit driving another pile, so the contractor built a wooden box around the pile and sealed the bottom of the box. At press time, he was still deciding whether to insulate the pile form or heat the water in the box.

Q. *How high can concrete be pumped vertically?*

A. It depends on the maximum placing pressure of the pump. As a rule of thumb, about 1.2 psi per foot of height is required for normal-weight, hard-rock concrete. Single pumps have been used since the early 1980s to place concrete over 1,000 feet high. Manufacturers say it's feasible to place concrete about 1,500 feet high with some of today's pumping equipment.

Q. *The late winter and early spring concreting seasons are coming up. Are there any special precautions we should take?*

A. We asked Art King of Materials Service Corp., Chicago, to help us answer your question. He recommends that you work with your supplier to get the most from your investment in admixtures and energy costs. When required, use heaters to warm up the forms or subgrade. Consider energy, labor, and materials costs to determine the most cost-effective approach. Although relatively expensive, nonchloride accelerators are available for work where chloride-based accelerators are prohibited. This may require an additional submittal to the architect/engineer.

As a general guide, if you're not comfortable laying on the subgrade, it may be too cold for the concrete. Consult the latest edition of "Cold Weather Concreting," ACI 306R, for recommended curing temperatures and durations. Art mentioned that many ready mix suppliers stop heating materials and rely on heated water as the weather improves. Check with your producer and review your materials and procedures if you are placing concrete during a cold snap.

Q. *We always have a difficult time hiring good shotcrete nozzlemen. When interviewing nozzlemen, how can we screen those who can shoot well from those who can only talk about shooting?*

A. Screening and hiring proficient workers can be difficult and time-consuming. Since field workers affect a project's overall quality, it's wise to devote some time to the hiring process. Also, with the costs of paperwork and training for each new hire, it's best to select a good candidate the first time out.

Consider using the ACI Committee 506 "Guide to Certification of Shotcrete Nozzlemen" for the screening process. This document suggests having each candidate take a written test and demonstrate workmanship. To quickly narrow the field of applicants, have each one answer the following 10 true/false questions taken from ACI 506. For a copy of the guide, contact the American Concrete Institute, P.O. Box 9094, Farmington Hills, Mich. 48333 (248-848-3800).

Quick Screening Quiz for Shotcrete Nozzlemen
Place T (true) or F (false) in the blank after each statement.
1. Dry-mix fine-aggregate shotcrete is also known as gunite. ____
2. When enclosing reinforcing steel, the nozzleman should:
    (a) Hold nozzle closer than usual. ____
    (b) In dry-mix application, slightly reduce the amount of water entering at the hose. ____
    (c) Keep face of bar clean so the steel can be seen until buried. ____
3. Shotcrete will easily bond to a concrete surface that has been cured with a regular concrete curing compound. ____

4. Shotcrete rebound should not be salvaged and worked into later batches of materials. ___

5. In the dry-mix process, a one-to-two mix proportion of cement to sand is considered a lean mix. ___

6. When water curing is specified, both wet- and dry-mix shotcrete surfaces should be kept moist for a minimum of seven days. ___

7. Curing fresh concrete is necessary to:
    (a) Keep surface clean. ___
    (b) Minimize shrinkage cracking. ___
    (c) Make concrete strong by chemical action. ___

8. Hose plugs can be caused by:
    (a) Wet sand. ___
    (b) Rocks or caked cement lumps. ___
    (c) Poor operation by gunman. ___
    (d) Too much water at nozzle. ___
    (e) Not blowing out hose when gun is shut down. ___

9. Sand pockets are caused by:
    (a) Not curing fresh shotcrete. __
    (b) Slugs from nozzle. ___
    (c) Hot sun and wind. ___
    (d) Holding nozzle too far from reinforcing steel. ___
    (e) Shooting at too much of an angle to the wall. ___
    (f) Not shooting into corners first. ___
    (g) Shooting over rebound. ___

10. It is possible for properly spaced reinforcing steel to be moved away from the shooting surface during shotcrete application if the steel is not securely tied back to the shooting surface. ___

Answers:
  1. T
  2. (a)T   (b)F   (c)T
  3. F
  4. T
  5. F
  6. T
  7. (a)F   (b)T   (c)T
  8. (a)T   (b)T   (c)T   (d)F   (e)T
  9. (a)F   (b)T   (c)T   (d)T   (e)T   (f)T   (g)T
 10. T

Vapor retarder under sidewalks?

Q. *For sidewalks placed directly on the subgrade soil, is a sheet-plastic vapor retarder useful or harmful?*

A. In temperate climates, we see no advantage to placing a vapor retarder beneath a sidewalk. If you're in a cold climate and the subgrade soil is poorly drained, a vapor barrier might be marginally useful. It could protect the concrete from freeze-thaw damage by reducing the amount of water it absorbs from the wet subgrade. But it might be harmful to seal this concrete with a nonbreathing sealer, thus trapping moisture within the concrete pores.

Q. *Will immersion-type vibrators damage epoxy-coated rebars? I've heard that they do. If so, how do you consolidate concrete that contains epoxy-coated bars?*

A. Steel-head vibrators might nick the epoxy coating. Because of this possibility some specifiers require use of a vibrator with a nonmetallic head. Coated vibrator heads are available from several manufacturers.

Q. *While we were getting ready to build some slabs that required protection against moisture intrusion, we searched for all the information we could find on vapor barriers. Looking through the American Concrete Institute booklet* Slabs on Grade *(2nd ed.), we noticed slip sheets mentioned in the same section that discusses vapor barriers. What are slip sheets? And do we need them too?*

A. A slip sheet is a layer of material placed between the slab and the subgrade to reduce the friction between them. When a concrete slab dries, it shrinks. As the concrete begins to shorten, some cracking can occur because the subgrade resists slab movement. Slip sheets reduce this frictional resistance and help reduce cracking. The greatest friction reduction is obtained when two sheets of plastic are placed on top of the subgrade directly beneath the slab. Puncturing the sheets before placing concrete allows water to escape from the bottom of the newly placed slab.

Slip sheets might be used under slabs with extra-long joint spacings or under post-tensioned slabs. A force applied to post-tensioning tendons induces a compressive prestress in the concrete, slightly shortening the slab. However, because of frictional resistance to this shortening, especially in long slabs, all the force in the tendons isn't transferred to the concrete. Thus, the compressive prestress may not be as great as desired. Slip sheets reduce friction so more of the tendon force is transferred to the concrete.

Don't confuse slip sheets with vapor barriers (or vapor retarders) just because they may be made of the same material. Unlike vapor barriers, which shouldn't be punctured, slip sheets are usually punctured purposely to allow water to escape from the bottom of the slab into the subgrade, and there is never a layer of granular material (blotter layer) on top of them. However, if slip sheets must also serve as vapor barriers, they aren't punctured.

Q. *I'm a remodeling contractor who builds a lot of additions, usually at the rear of an existing house. We pour our own footings, foundations, and flatwork, but often the 10-cubic-yard ready-mixed concrete trucks are too heavy or too big to get to the home's backyard without damaging the lawn and landscaping. Placing concrete by wheelbarrow is slow and labor-intensive. What's the most economical way to solve this problem?*

A. Some concrete producers have smaller trucks that carry 5-cubic-yard loads and are more maneuverable, so they are better able to fit between trees and shrubs. Putting plywood sheets in a truck's wheel path also reduces lawn damage.

Walk-behind or ride-on power buggies are another option. Most have capacities from 9 to 16 cubic feet, but larger units are also available. If you own a skid-steer loader, consider using a concrete placing attachment. (See *Concrete Construction*, May 1997, page 436-442 for descriptions of several of these units.)

Small-line concrete pumps can also be used successfully for pours where access is a problem. If you do a lot of this kind of work, buying a small-line pump may be more economical than using a pumping service. Boom pumps—with the boom extended over the house and into the backyard—have been used to place concrete made with ¾- or 1-inch maximum-size aggregate. Higher pumping costs are the major disadvantage of using the larger boom pumps.

Granular bases under slabs on grade

Q. *What purpose is served by placing a granular base under an interior concrete slab? Why not just place concrete directly on the compacted subgrade?*

A. If the subgrade is uniform (no soft spots) and a vapor barrier is placed on it, you probably don't need an aggregate concrete base course. A base serves two main purposes:
- It provides uniform support for the slab
- It serves as a capillary cutoff that prevents moisture from wicking up through the subgrade and collecting beneath the slab

A granular base also may be easier to fine-grade before placing the concrete. If the subgrade is difficult to grade, slab thickness may vary.

Building post-tensioned slabs on grade

Q. *We're going to be placing and finishing the concrete for a post-tensioned concrete slab on grade in a warehouse. It's the first time we've done a post-tensioned floor. How does a post-tensioned floor differ from a conventional floor, and where can we get information about construction methods for this kind of work? We want to avoid problems so we're not learning as we go at the jobsite.*

A. In a post-tensioned slab on grade, a grid of high-strength post-tensioning tendons re-places the welded wire fabric or conventional rebar normally used. For a lightly loaded in-dustrial floor, the slab is placed 4 inches thick with tendons in both directions at 2- to 5-foot intervals.

After the concrete has reached sufficient strength, the tendons are stressed by hydraulic jacks to an effective force of about 25,000 pounds. This force is permanently transferred from the tendons to the concrete through anchorage devices at the ends of the tendons. This pro-duces an internal compressive force in the concrete that makes the floor more resistant to cracking under load or cracking due to concrete shrinkage or temperature variations. Other benefits of post-tensioning include elimination of most joints, reduced slab thickness, and re-duced excavation costs.

You can get specific information about post-tensioning in the 1998 publication *Construction and Maintenance Procedures Manual for Post-Tensioned Slabs-on-Ground*. It is available from the Post-Tensioning Institute, 1717 W. Northern Ave., Suite 114, Phoenix, AZ 85021. Also, there's an article on post-tensioned floors in the January 1987 issue of *Concrete Construction* magazine, and an article on post-tensioned residential foundations in the February 1991 issue of *Concrete International* published by the American Concrete Institute.

Find out who the post-tensioning subcontractor will be for the project. Ask them to take you to another one of their jobs already in progress. By touring a job and asking what's most likely to go wrong, you'll have a head start in avoiding most of the common problems.

Vibrating screeds and air entrainment

Q. *Doesn't a vibrating screed cause a loss of entrained air at the surface of the concrete, right where it is needed most?*

A. When vibration is done properly, it removes mainly large air bubbles that don't contribute to the freeze-thaw durability of the surface. The air void spacing factor, one of the most critical aspects of air-entrained concrete, remains relatively constant even though the average size of air voids gets smaller. Prolonged vibration may harm the air void system. To avoid this, instruct contractors not to leave vibrators running when the screed isn't being moved forward and not to make multiple passes over the same concrete with the screed.

Q. *A customer asked our field rep whether there is an allowable time limit for successive concrete placements during a monolithic pour. I think the customer was asking this in reference to cold joints.*

A. The American Concrete Institute (ACI) literature on building codes or structural concrete specifications provide no reference to a set time limit. That's probably because time isn't the only factor to consider in determining initial set. In cool, damp weather you can wait longer to place the next batch than you can during hot, dry weather. As long as the previously placed concrete is still plastic (an internal vibrator will penetrate it), you can safely place the next layer of concrete without producing a cold joint.

ACI's "Guide for Consolidation of Concrete," ACI 309R-96, gives the following advice: "To avoid cold joints, placing should be resumed before the surface hardens. For unusually long delays during concreting, the concrete should be kept live by periodically re-vibrating it. Concrete should be vibrated at approximately 15-minute intervals or less depending upon job conditions. However, concrete should not be overvibrated to the point of causing segregation. Furthermore, should the concrete approach time of initial setting, vibration should be discontinued and the concrete should be allowed to harden. A cold joint will result and suitable surface preparation measures should be applied."

Q. *This past winter, a contractor in our area placed several concrete slabs on frozen subgrade. What problems can result from this practice?*

A. There are several good reasons for not placing concrete on a frozen subgrade:

1. Some soils expand or heave when they freeze. Later, when they thaw, the ground can subside and the slab will lose its subgrade support. This can cause the concrete to crack.

2. Cold concrete at the bottom of the slab will take longer to set than concrete near the surface. When this happens, the surface may crust over and look as though it is ready for finishing. The rest of the concrete, however, may not be firm enough to support the weight of a finisher or a power trowel. This condition can cause concrete to bulge out underneath a power trowel and create a wavy surface. In addition, the surface may crack.

3. The cold subgrade may actually freeze a portion of the concrete. Ultimate strength reductions of up to 50% have been reported in concrete that has frozen within a few hours of placement or before it attains a compressive strength of 500 psi.

Q. *We recently experienced some problems on a bridge deck repair project that required partial- and full-depth patches. Two months after completion, the repairs were inspected, and the surface of many of the patches was worn and scaled. Compressive strength testing was done on cores taken from the patches, and some of the breaks were low. Throughout the project, heavy traffic was allowed to continue on the bridge. Could vibration of the deck caused by allowing traffic to continue have affected the concrete quality?*

A. The effect of continuous vibrations on concrete quality was reported in "Traffic-induced Vibrations and Bridge Deck Repairs," *Concrete International*, May 1986. The study concluded that traffic-induced vibrations appear to have no detrimental effect on compressive strength in bridge deck repairs if high-quality, low-slump concrete is used. In fact, compressive strength appears to increase slightly for low-slump concretes when vibrated. However, for concrete having slumps greater than about 4 inches, the study found that vibration can be detrimental to compressive strength. The study suggested that higher-slump concrete will have significantly more bleedwater, which will rise during vibration. Therefore, high-slump concrete test specimens will have a layer of high-water-cement-ratio, low-strength concrete at the upper end of the cylinder. During the compression strength testing, this weaker concrete should dominate the cylinder strength.

Wet- or dry-process shotcrete for spillway repairs?

Q. *Our company will be repairing a concrete spillway for a dam. We'll be repairing slab-on-grade work and walls. We're investigating several repair methods including wet- and dry-process shotcreting. What are the relative advantages and disadvantages of these methods?*

A. According to Rusty Morgan of Hardy BBT Ltd., economics usually dictates the method used. Wet shotcreting requires continuous shooting and is well suited for high-production jobs. To use this method economically for repair you have to do much of the repair surface preparation ahead of time. But on spillway repairs, this may require lots of scaffolding and drive up costs. Dry-mix methods are more commonly used because they work better when there's frequent starting and stopping. You can use less scaffolding because each area to be repaired can be done from surface preparation to shooting before moving on to the next area. Some contractors use a high-reach crane instead of scaffolding, but the dry process works well with this approach too.

Removing water from footing forms

Q. *One of the residential contractors our ready mix company works with has the habit of placing concrete in footings that contain pools of water. The contractor claims that the concrete simply displaces the water and sees no problem with this technique. Is this an accepted practice, or should the water be removed before placing concrete?*

A. American Concrete Institute Committee 332, "Guide to Residential Cast-in-Place Concrete Construction," offers the following recommendation: "Pools of rainwater that have collected in the footing forms must be pumped out, and all water that has collected in forms or on the grade should be removed before placing concrete. It is not always possible to get the surface completely dry, particularly where the water table is high. If so, the concrete should be placed in a manner that displaces the water without mixing it into the concrete."

Concrete set time during cold temperatures

Q. *We receive questions from our ready mix customers concerning accelerating the set of the concrete in periods of cold weather. We have used accelerators and Type III cement during the winter months, but we regularly heat the mix water from fall to spring. How will hot mix water affect the setting time of the concrete? Are there rules-of-thumb that we can tell the contractors?*

A. As concrete temperature decreases, set time increases. To increase concrete temperature in cold weather, add hot water or heated aggregates to the mix. In relatively mild climates, or when temperatures commonly do not fall below freezing, it is sufficient to heat the mixing water alone. As a general rule, each 5-degree increase in water temperature will increase concrete temperatures by 1 degree. Use the figure below for a more exact calculation of the required mix water temperature to produce heated concrete. To use the graph, first you must calculate a weighted averaged temperature of the aggregates and cement as follows:

$$\text{Weighted Average Temp.} = \frac{(\text{Temp. of Aggregates} \times \text{Wt. of Aggregates}) + (\text{Temp. of Cement} \times \text{Wt. of Cement})}{\text{Wt. of Aggregates and Cement}}$$

Once this temperature is known, enter the graph on the horizontal axis and move vertically until intersecting the desired concrete temperature. Once at the desired concrete temperature, move left to find the mixing water temperature required. The graph is based on the following equation that can also be used to calculate resulting concrete temperature with the addition of hot water:

$$T = \frac{0.22(T_a W_a + T_c W_c) + T_w W_w + T_{wa} W_{wa}}{0.22(W_a + W_c) + W_w + W_{wa}}$$

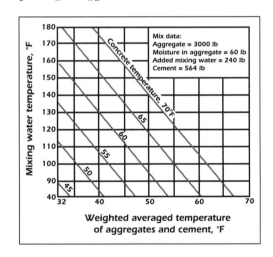

where
T = temperature of the freshly mixed concrete
$T_a, T_c, T_w,$ and T_{wa} = temperature of aggregates, cement, added mixing water, and free moisture on aggregates, repectively
$W_a, W_c, W_w,$ and W_{wa} = weight of aggregates, cement, added mixing water, and free moisture on aggregates, respectively
In a plant where the added water is made by batching hot and cold water to obtain the desired concrete temperature, use the procedure from the following example:
1. Obtain temperatures of hot and cold water.
Hot water temperature: 170° F
Cold water temperature: 50° F
Desired concrete temperature: 70° F
2. Run a trial batch with cold or hot water alone and measure concrete temperature.
Added 200 pounds per cubic yard at 170° F. The resulting concrete temperature was 85° F.
3. Calculate desired increase or decrease in concrete temperature.
The desired concrete temperature was 70° F and the concrete produced was 85° F. A decrease of 15° F is desired.

4. Use the figure below to calculate the quantity of water to be replaced. Add hot water to raise the temperature and cold water to decrease the temperature.

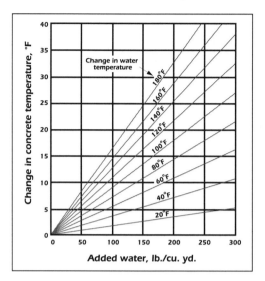

The difference in water temperature between hot and cold is 170° − 50° = 120° F. Enter the graph at 15° desired change in concrete temperature on the vertical axis. Move to the right until intersecting the change in water temperature of 120° F and move down to find the amount of hot water to be replaced by cold water. The hot water added should be replaced with 140 pounds of cold water to reach the desired temperature.

Do not place hot water in direct contact with cement. This may lead to flash set of the cement. Avoid adding water hotter than 180° F and mix cement with aggregates before adding heated water to the mix.

A rule-of-thumb is set time increases by about one-third for each 10-degree drop in concrete temperature. The table below shows the effect of concrete temperature on set time.

How concrete temperature affects set time

Concrete temperature,°F	Set time, hours
70	6
60	8
50	11
40	14
30	19

Source: "Cold-weather finishing," *Concrete Construction*, November 1993

Chapter 18

Production

Q. *How often should I clean my strand chucks?*

A. Strand chucks should be cleaned and inspected after every use. It is a recommended practice to have enough chucks on hand to keep one set in the chuck cleaning room and one set to use on your beds. This way, your chuck cleaning person can send cleaned and inspected chucks to the yard and release, clean, and inspect the chucks that were used the previous day.

Strand chucks require cleaning

Q. *I am interested in expanding my precasting business, possibly with a line of retaining wall products. What are some of the markets I should explore? Can you offer any advice?*

A. There are two good reasons to consider adding a new product to your line. You can extend your product line if you have extra capacity to provide new products to your current customers, or you can replace an existing commodity product with an engineered product that has a higher profit margin. Before adding any new product to your line, ask two questions:
- Who are my customers?
- What are my strengths?

A retaining wall is an engineered product that will need backup, including some design. At times, a consultant will show a retaining wall on plan as a straight line, leaving it up to the contractor to choose the retaining wall to be used.

The 1980s saw an increase in retaining wall construction. This growth was fueled by limited rights-of-way and the rising cost of land. In addition, the growth of the retaining wall market was due to the precast advantage. Precast walls take less time to build, cause less disruption to local activities, and cost less than cast-in-place walls. Retaining wall construction can be divided into three major markets: highway, industrial/commercial, and landscape/residential.

Deciding which market to be in is a function of your customer base and the retaining wall product you choose to produce. For example, if you sell landscape items, a block system would fit your customer base. If you sell box culverts and bridge beams, a precast system that meets public agency requirements would fit well with your marketing efforts.

Successful sales of precast retaining walls mean selling a solution to a specific problem, which is a different ball game than selling commodity products. It includes selling the availability of solutions (marketing and advertising), the technical capability of the solution (engineering), the means to implement the solution (precast concrete), and assistance in the actual implementation of the solution (construction assistance).

The system best suited to you is a function of the market you work in and the services you now provide. If all you would like to do is pour concrete and provide transportation, you can be a subcontractor to a company providing a complete retaining wall package of marketing, engineering, materials supply, and construction assistance.

Adding a new product to your line

On the other hand, if you will do the marketing and have salespeople who will learn about retaining walls and can develop a project, then you can work through a licensing arrangement that provides the system and services you do not have in house.

Separating storm and process water

Q. *The county engineer suggested we revise the drainage around our plant to separate our storm water from our process water. Why would I want to do this, and how would this affect my storm water permit?*

A. Storm water refers to rain and melted snow that falls onto, and discharges off, plant property. When the precipitation falls onto non-plant-affected areas of the site, or those areas that do not contain source materials (materials which could conceivably contribute pollutants to storm water), many regulatory agencies do not require any specialized management or treatment of the runoff produced. When runoff flows across plant areas or contacts source materials, producers are usually required by their storm water permit to manage via "best management practices" (BMPs) or otherwise treat the runoff. Should the runoff become commingled with process water generated at the plant prior to discharge, that runoff must then be treated

Figure 1. This dam separates storm water runoff from plant process water.

Figure 2. This three-chamber washout pit's size was minimized by separating storm water runoff from process water.

as process water, which usually has considerably stricter discharge requirements. Keeping storm water separate from process water can reduce the severity of the discharge requirements. Should your permit require retention or detention of process water prior to discharge, separating storm water from process water can also reduce the size and volume of retention or detention areas. (Retention areas are impoundments that retain storm water for more than 24 hours. Detention areas do not have any permanent standing water, as they temporarily detain a portion of storm water runoff for up to 24 hours after a storm.)

Under most storm water permits, producers are required to implement BMPs for their Storm Water Pollution Prevention Plans. The separation of storm water runoff from process water flow is a recognized BMP. Drainage ditches, curbs, swales, and kneewalls are structures that keep runoff water away from process areas.

Many plants construct process drainage that keeps storm water runoff from stockpiles, scales, and haulway areas separate from concrete-processing areas. Since most of this water may be high in total suspended solids, remediation may only require a settlement area. This water often becomes a source of mixing water.

Water from the concrete production area may have a pH level near 12.0 due to the concrete rinse water. Prior to discharge, the pH must be lowered. Limiting the amount of water in this treatment area helps reduce operating costs. Examples of good storm water management practices were submitted by CAMAS Colorado Inc.'s Flanagan Ready Mix plant in Broomfield, Colo., as part of an entry in the National Ready Mixed Concrete Association's Environmental Excellence Award 1996 contest.

Figure 1 shows a check dam in the plant's storm water swale. The swale extends through the center of the property and directs storm water into a detention area where settlement is allowed to occur. Process water is directed into a retention pond and not allowed to commingle with storm water. Figure 2 shows a three-chambered washout pit for the plant area. Without the separation of storm water runoff, the washout pit area would have to be substantially larger.

Reference

Environmental Management Practices, Publication No. 191, National Ready Mixed Concrete Association, 1997.

Removing moisture from compressed air lines

Q. *We get low productive efficiency in air-operated equipment and controls because of excessive air-line water. How can we eliminate the moisture?*

A. Most air compressors are equipped with aftercoolers designed to lower the compressed air's temperature and remove some water vapor from the airstream. But an aftercooler can't remove all the moisture contained in ambient air.

For instance, a 25-horsepower compressor delivering 100 cfm at 100 psig can produce 18 gallons of water per day at fairly standard conditions of 90° F ambient temperature and 50% relative humidity. An aftercooler will remove approximately 75% of this moisture, leaving about 6.2 gallons of water per day to run through the air-supply system. Engineers suggest two ways to remove this moisture.

One way is to further lower the compressed air's temperature after it's passed through the aftercooler, causing water vapors to condense. Refrigerated and thermal mass dryers cool the compressed air as it flows through refrigeration circuits. However, the refrigeration systems don't work well under dusty or below-freezing conditions.

A better choice for concrete producers is the second dryer style, which employs desiccants to remove moisture. This type of dryer is more common in concrete production plants because it is effective in winter weather, when the pipeline temperature drops below 35° F. Manufacturers offer single- and double-tower desiccant dryers that remove moisture differently.

A dessicant air dryer can remove excessive moisture from compressive air lines.

In a single tower dryer, *absorbent* desiccant beads attract vapor. Because the beads dissolve slowly, they require occasional replenishment. In a twin-tower style dryer, water is captured with *adsorbent* desiccants. Rather than dissolving upon contact, these desiccants hold the water until it is purged by either internal heaters or countercurrent airflow. Since the tower must be removed from service during purging, each dryer has two desiccant-filled towers. While one is drying compressed air, the other is being regenerated for the next drying cycle.

Managers with plants using rotary vane compressors usually avoid twin-tower dryers because airstreams from those compressors often contain higher-than-normal quantities of oil. If this oil is not removed, it will damage the desiccant and cause other hazards.

Producers interested in reducing airstream water should consult air dryer manufacturers for help in sizing and designing their systems.

Reference

Frederick Sitter, "Selecting an air dryer for dust-collection systems," *Concrete Journal*, October 1995, pp. 806-809.

A possible use for reclaimer sludge

Q. *Can concrete reclaimer sludge be used to reduce soil acidity?*

A. If preliminary results from a University of Wisconsin Soil and Forage Analysis lab study are correct, highly alkaline reclaimer sludge may provide soil nutrients some farmers need. The study, led by John B. Peters, the lab's director, focused on raising the pH of soil planted in alfalfa, a very acid-sensitive crop.

"Concrete sludge performed at least as well as conventional aglime in raising the pH levels and increasing the total alfalfa dry material yield in the greenhouse study," says Peters.

The pH change occurred relatively soon after application, probably due to the slurry's fineness. When concrete reclaimer sludge was applied at a 0.5 calculated lime rate, soil pH changes were very similar to those for conventional aglime using a 1.0 rate.

The Wisconsin Ready Mixed Concrete Association (WRMCA) is currently working with the state's Department of Natural Resources to approve reclaimed slurry as a liming agent. More field studies are anticipated in the spring. For more information about the study, contact WRMCA at 414-529-5077.

Q. *Our floating pumps occasionally suck up trash and vegetation from the lake's bottom. How can we protect our water flowmeter from the debris that gets through the pump's strainer?*

Protect your flowmeter

A. Intake strainers on large-volume freshwater pumps provide protection for the pump vanes without minimizing water flow. While these prevent large pieces of debris from entering the pipe, they are not designed to completely clean the water. Particles large enough to damage turbine flowmeters and affect concrete quality often pass through.

Since blade and bearing designs vary widely in water-meter design, it is important for producers to follow manufacturer installation recommendations. The mesh size of the strainer's screen is based on the size of the flowmeter. The table below lists recommended strainer mesh numbers.

Typical mesh numbers for upstream strainers

Flowmeter size, inches	Mesh no.
½, ¾	100
1, 2, 3	80
4, 6	60
8, 10, 12	40

Source: Flow Measurement Engineering Handbook, R.W. Miller, McGraw-Hill Publishing Co.

Bronze-bodied "Y" strainers are the best choices. You could use a less-expensive cast iron-bodied strainer, but over time it can rust and corrode. The strainer should be the same diameter as the meter's intake pipe to avoid flow restriction.

When installed too close to the meter's measuring turbine, strainers compromise the accuracy of flow readings. This is caused by cavitation and turbulence created as water passes through the strainer's screen openings. Manufacturers recommend a minimum distance between strainer and meter inlet of 10 times the pipe's inside diameter. For example, a strainer on a 3-inch pipe should be installed at least 30 inches away from the meter.

Also, when installing the strainer, make sure no electrical equipment is in the strainer-drain plug's outfall path.

Water-meter manufacturers suggest that all concrete plants, even those supplied with municipal water, benefit from properly placed strainers. Producers have reported water-meter damage by debris from pipeline repairs performed miles away from the plant, or scaling from the linings of older city pipes.

Q. *Contractors often request us to provide a pumpable mix. When the mix proportions aren't specified, what should we consider when designing a mix to be pumped?*

A. A pumpable mix must slide along the wall of the pipeline and be able to flow around bends. The mortar acts as the lubricant, so concrete mixes for pumping must be plastic, not harsh. Give particular attention to the mortar and the amount and size of coarse aggregate.

Pumping is sensitive to variations in gradations and in batch proportions. After you develop a pumpable mix, quality assurance is needed to ensure repetition of all the factors that produce the mix.

Coarse aggregate. The amount of coarse aggregate used in the mix depends on the shape and size of the aggregate. The maximum size of an angular coarse aggregate is limited to 33% of the smallest inside diameter of the pump or pipeline. For well-rounded coarse aggregate, the maximum size can be up to 40% of pipe diameter.

As with most aggregates used in concrete, coarse aggregate must meet ASTM C 33. Preferably, the gradation will be as close to the middle of the range as possible. However, grading uniformity on a daily basis is most important.

The maximum size of the coarse aggregate has a significant effect on the amount that can be used. The quantity of aggregate per unit volume must be reduced as the maximum size is reduced. For example, ACI 211.1, "Standard Practice for Selecting Proportions for Normal, Heavyweight, and Mass Concrete," recommends using 69% of 1-inch-diameter coarse aggregate versus only 48% of 3/8-inch-diameter coarse aggregate when using fine aggregate with a fineness modulus of 2.60. ACI 211 notes that this quantity of coarse aggregate may have to be reduced up to 10% for more workable concrete, which may be required when pumping.

When absorptive aggregates are used, especially lightweight materials, pumping pressure may force mixing water into aggregate pores. The resulting slump loss reduces pumpability. Presoak the aggregate stockpiles for at least three days to minimize this problem. Turning the pile over with a front end loader helps to ensure uniform presoaking.

Fine aggregate. Fine aggregate, cement, and water create the paste that lubricates the coarse aggregate and holds it in suspension, making the mix pumpable. Fine aggregate properties have a greater effect on pumpability than do coarse aggregate properties.

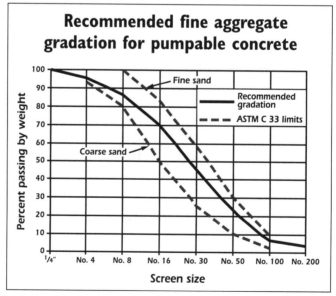

The solid line represents an ideal fine aggregate gradation for concrete that will be pumped. When available materials don't have this gradation, fine sands (gradations near the upper dashed line) are preferred over coarse sands.

Fine aggregate gradation should also conform to ASTM C 33 requirements. ACI 304.2R, "Placing Concrete by Pumping Methods," suggests that particular attention should be given to those proportions passing the finer screen sizes. For pumplines less than 5 inches in diameter, 15% to 30% should pass the No. 50 screen and 5% to 10% should pass the No. 100. If the smaller sizes are missing, small quantities of materials such as crusher dust, wash pit settlings, fly ash, and beach or dune sand will often improve pumpability. To meet ASTM C 33, the fineness modulus of the sand should be between 2.30 and 3.10. Concretes made with fine sands (lower fineness modulus) usually pump better than those made with coarse sands, although very fine sands increase the water and cement requirements and may produce sticky mixes that increase the pressure needed to pump.

The gradation recommended in the graph is a guide to selecting a suitable fine aggregate. The upper and lower limits of ASTM C 33 are shown with dotted lines, and the solid line represents the recommended gradation. This gradation meets the requirements of ASTM C 33 and the percentages passing the No. 50 and No. 100 screens are within the limits previously stated. In practice it may be difficult to duplicate this gradation; however, a gradation near the upper limit will be more desirable than one towards the lower. Low-cement-content concrete made with coarse sands, or concrete made with gap-graded sands, bleed excessively and are likely to cause line blockages.

Water and slump. Establish the optimum slump and maintain this slump throughout the pumping portion of the job. Slumps below 2 inches and above 6 inches may cause pumping problems. Depending on the pump used, mixtures with low slumps may not feed well from the hopper into the pumping chamber. When high slumps are produced by adding water only, segregation may cause pumpline blockages.

Cement content. Choosing the cement content for pumpable concrete follows the same principles used for any concrete. Don't fall into the trap of using extra cement as the only cure for pumping deficiencies. This is both uneconomical and shortsighted. Instead, pay close attention to the coarse and fine aggregate gradations.

Admixtures. Many admixtures will improve pumpability. Some of the admixtures commonly used are:
- Water reducers, including superplasticizers
- Air-entraining agents
- Pozzolans

A full-scale test to ensure pumpability is always recommended when designing a pump mix. During a test, duplicate job conditions as closely as possible.

Reference

ACI 304.2R, "Placing Concrete by Pumping Methods," American Concrete Institute, Farmington Hills, Mich.

Adding water to concrete

Q. *Is there a rule of thumb about the effects of adding water to a concrete mix?*

A. The basic rule of thumb is that adding 1 gallon of water per cubic yard increases concrete slump by about 1 inch. Because of the more fluid mix, air content also is likely to increase by 0.5% to 1%. Concrete compressive strength decreases because the increased water-cement ratio produces a weaker paste and the increased air volume means there is less solid material present. The compressive strength decrease could be 200 to 300 psi, considering both factors. In addition, shrinkage potential increases by about 10% and as much as ¼ bag of cement is wasted.

Keep in mind that rules of thumb should be used only for quick evaluations of concrete mix changes and that they may not apply in all situations. Some other rules of thumb that may be helpful are:

If fresh concrete temperature increases 10° F . . .
- About 1 gallon of water per cubic yard maintains equal slumps
- Air content decreases about 1%
- Compressive strength decreases about 150 to 200 psi

If the air content of fresh concrete. . .
- Increases 1%, then compressive strength decreases about 5%
- Decreases 1%, then yield decreases about ¼ cubic foot per cubic yard
- Decreases 1%, then durability decreases about 10%

How many vibrators are needed?

Q. *We are having trouble producing smooth, defect-free, dense concrete manholes. If we increase vibration times, these problems disappear, but vibration then takes longer than usual. Before mounting extra vibrators, I want to find a standard way to calculate the amount of external vibrators needed for this application.*

A. *The Vibco Handbook & Equipment Guide for External Concrete Vibrators* offers the following guidelines:

1. Total the weight of the form and the fresh concrete. This is equal to the vibrator force needed. For example, if the concrete and the form for the manhole weigh 1,500 pounds, you need 1,500 pounds of vibrator force.

2. This force should be adjusted for low-slump concretes. For slumps 3 inches or greater, no adjustment is needed. For slumps in the range of 1 to 2 inches, 75% more vibrator force is needed. For zero-slump concretes, as much as 200% more vibrator force is required to consolidate the concrete. The manufacturer can supply you with the vibrator force produced by each model.

3. The attachment bracket may not transfer all of the force generated. Only brackets that secure the vibrator firmly will transfer 100% of the force. If you can't increase the rigidity of the bracket, use a higher vibrator force, increase vibration time, or use more vibrators to provide adequate consolidation.

4. Vibrate the manhole until no more air bubbles break on the surface and the surface glistens. Vibration can be done with one vibrator and two brackets. Mount one bracket low and one high. Move the vibrator from the lower position to the higher position once the concrete within the form reaches the elevation of the higher bracket.

Using fibers for GFRC panels

Q. *What type of fibers are used in glass-fiber reinforced concrete (GFRC) panels? Also, in what quantity are the fibers used?*

A. According to *Recommended Practice for Glass Fiber Reinforced Concrete Panels*, published by the Prestressed/Precast Concrete Institute, only high zirconia (minimum 16%) alkali-resistant glass fibers specifically designed for alkali resistance should be used in GFRC panels. A glass fiber content of 5% by weight is optimum for GFRC mix designs. Lower fiber contents result in lower early ultimate strengths, while higher fiber contents can lead to composite compaction and consolidation problems.

Determining bar tie size

Q. *We are adding several new products to our precast product line and the bar sizes are different than we usually use. What is an easy way to determine the bar tie size needed to tie two reinforcing bars?*

A. Most bar tie manufacturers have a chart which can be used to determine the bar tie size. You select the bar size from the vertical column and then follow that row to the other size bar to be tied and the chart will give you the recommended size.

Dick Taylor, Florida Wire and Cable Co., Jacksonville, Fla.

Bar Tie Table

Wire Ties are measured length overall. This table will help in estimating tie requirements for reinforcing bars

Rod Sizes	No. 2 1/2	No. 3 3/8	No. 4 1/2	No. 5 5/8	No. 6 3/4	No. 7 7/8	No. 8 1	No. 9 1-1/8	No. 10 1-1/4
No. 2 1/4	3-1/2	4	4-1/2	5	5-1/2	6-1/2	7	7	7-1/2
No. 3 3/8	4	4-1/2	5	5	5-1/2	6-1/2	7	7	7-1/2
No. 4 1/2	4-1/2	5	5	5-1/2	6	6-1/2	7-1/2	7-1/2	8
No. 5 5/8	5	5	5-1/2	6	6-1/2	7	8	8	8-1/2
No. 6 3/4	5-1/2	5-1/2	6	6-1/2	6-1/2	7-1/2	8	8-1/2	8-1/2
No. 7 7/8	6-1/2	6-1/2	6-1/2	7	7-1/2	7-1/2	8-1/2	9	9-1/2
No. 8 1	7	7	7-1/2	8	8	8-1/2	9	9-1/2	10
No. 9 1-1/8	7	7	7-1/2	8	8-1/2	9	9-1/2	10	10-1/2
No. 10 1-1/4	7-1/2	7-1/2	8	8-1/2	8-1/2	9-1/2	10	10-1/2	10-1/2

Q. *Our ready mix plant is planning on using a new source of sand. What are some of the requirements for the fine aggregate gradation? Do our mix designs need to be redesigned after switching to a new source of sand?*

Grading requirements of sand

A. Sands used in the production of concrete should meet the requirements set forth in ASTM C 33, Standard Specification for Concrete Aggregates. For sands used in concrete, the fineness modulus can range from 2.3 to 3.1. The coarser the sand, the higher the fineness modulus. The fine aggregate gradation should also be within the following limits:

Sieve	Percent passing
⅜ inch	100
No. 4	95 to 100
No. 8	80 to 100
No. 16	50 to 85
No. 30	25 to 60
No. 50	10 to 30
No. 100	2 to 10

One of the most important characteristics of the fine aggregate grading is the amount of material passing the Nos. 50 and 100 sieves. Inadequate amounts of materials in this range can cause excessive bleeding, difficulties in pumping concrete, and difficulties in obtaining a smooth troweled finish. ASTM C 33 permits the lower limits for percent passing the Nos. 50 and 100 sieves to be reduced to 5 and 0 respectively, provided:

1. The aggregate is used in air-entrained concrete containing more than 425 pounds of cement per cubic yard and having an air content of more than 3%, or

2. More than 515 pounds of cement per cubic yard are used in non-air-entrained concrete, or

3. An approved mineral admixture is used to supply the deficiency in material passing these sieves.

Small variations in fine aggregate grading can affect the concrete workability. For this reason, ASTM C 33 states that for continuing shipments of fine aggregate from a given source, the fineness modulus should not vary by more than 0.20 from the fineness modulus assumed in selecting mix proportions. If this value is exceeded or a new source of fine aggregate is used, suitable adjustments must be made in the proportions of fine and coarse aggregate. It is often more economical to maintain uniformity in producing and handling aggregates than to adjust proportions for variations in grading.

Rust and strand performance

Q. *As we were about to start production of the prestressed members for a new job, an engineer from the testing lab inspected the steel strand. He rejected it because it was too rusty. Is there a way to determine whether rusting will have a harmful effect on strand performance?*

A. Determining an acceptable amount of rust on prestressing steel strand is difficult for the inspector and producer. ASTM A 416 states, "Slight rusting, provided it is not sufficient to cause pits visible to the unaided eye, shall not be cause for rejection." The real concern is with pitting. Pits are stress raisers, reducing the strand's ability to withstand repeated or fatigue loading.

Inspection is based upon the visible estimate of pitting in the strand. The PCI Manual suggests removing rust with a pencil eraser to expose any pitting. A more practical method is to use a Scotch Brite Cleaning Pad No. 96, which cleans a larger area for observation. After the strands are cleaned, a set of visual standards can be used to help field inspectors make proper judgments of strand suitability.

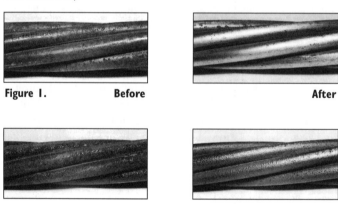

Figure 1.   **Before**   **After**

Figure 2.   **Before**   **After**

Figure 1 and Figure 2 show two different strand surfaces before and after pad cleaning. Pitting is visible in the after photo for Figure 2, so the strand should be rejected.

Reference

A.S. Sason, "Evaluation of Degree of Rusting on Prestressed Concrete Strand," *PCI Journal*, V. 37, No. 3, May-June 1992, pp. 25-30.

Surface texture of block

Q. *Our plant manufactures concrete block. On a few occasions we have had problems with the surface texture on the face shell. The texture is too coarse and contains many voids. What could be the cause of this?*

A. A coarse texture can be a problem for customers. If the block is to be left exposed or receive a finish such as paint, the different textures of the face will be noticeable. A well-known block machine manufacturer gave two causes for this problem: segregation and vibration.

Segregation may be due to the mix design. Also, a loss of some mix ingredients in the process of making the block may be the cause. Therefore, check the machine hopper for buildup of fines on the sides.

Also, vibration of the block may be incorrect or insufficient. Check the vibrator drive system and components, making sure all are properly adjusted and running at the correct speeds. Check the molds for loose parts and tighten or replace loose or worn parts. These adjustments should correct a rough textured block problem.

Q. *We manufacture concrete pavers and are experiencing a severe efflorescence problem. Can we reduce efflorescence by using admixtures such as plasticizers or pozzolans? Or should we use silicone-type water repellents or film-forming sealers on the finished product?*

A. Most efflorescence occurs when soluble calcium hydroxide, a byproduct of cement hydration, leaches out of the concrete and reacts with atmospheric carbon dioxide to form insoluble calcium carbonate. A German researcher (see reference) found that plasticizers help to reduce efflorescence by improving the compactibility of concrete used to make pavers. Less water penetrates a well-compacted concrete, and thus less dissolved calcium hydroxide is transported to the surface.

Pozzolans react with calcium hydroxide to form insoluble calcium silicate hydrates that also decrease concrete permeability. Because this reaction occurs at a slow rate, however, efflorescence can still occur. Also, an excessive amount of pozzolan would be needed to react with all of the calcium hydroxide in concrete.

Silicone-type water repellents increase the surface tension of water, thus preventing liquid water from penetrating the concrete. However, silicones don't prevent the passage of water vapor. Once this vapor condenses and fills capillary voids, calcium hydroxide can still rise to the surface, causing efflorescence.

The researcher found that a thin coat of an acrylic or other polymer-type sealer does seal the pores, preventing efflorescence. However, coatings can be applied only to pavers produced on a board (single-layer) machine, and only the top surface can be coated. If a cube of these coated pavers gets wet, the pore water can transport dissolved calcium hydroxide to the uncoated bottoms of the pavers. Calcium carbonate may then form on the tops of the underlying pavers.

You didn't mention the effect of cement-sand ratio on efflorescence. This, along with water-cement ratio, has a pronounced effect on paver porosity and efflorescence. We'll discuss this in a future issue of *The Concrete Producer*.

Reference
P. Kresse, "Efflorescence and Its Prevention," Betonwerk + Fertigteil-Technik, October 1991, pp. 73-87.

Q. *We had a discussion recently about the three weight classes of concrete block—lightweight, medium-weight, and normal-weight. Our question is this: At what point does a lightweight block become a medium-weight block, and a medium-weight block become a normal-weight block?*

A. According to the *Concrete Masonry Handbook*, published by the Portland Cement Association, the weight class of a concrete masonry unit is based on the density or oven-dry weight

Cures for paver efflorescence

Concrete block weight classes

per cubic foot of the concrete it contains. A unit is considered lightweight if it has a density of 105 pcf or less, medium-weight if it has a density between 105 and 125 pcf, and normal-weight if it has a density of more than 125 pcf. The density of concretes containing various aggregates is shown in the table.

Concrete unit weight ranges with various aggregates

Concrete	Unit weight, pcf
Sand and gravel concrete	130-145
Crushed stone and sand concrete	120-140
Air-cooled slag concrete	100-125
Coal cinder concrete	80-105
Expanded slag concrete	80-105
Pelletized fly-ash concrete	75-125
Scoria concrete	75-100
Expanded clay, shale, slate, and sintered fly ash concrete	75-90
Pumice concrete	60-85
Cellular concrete	25-44

Preventing sand streaking in precast products

Q. *We make precast septic tanks and often get water trails on the tank surfaces even though we use admixtures. Is there a way to avoid these, or will we always have some trails?*

A. The condition is called sand streaking and it's caused by excessive bleeding. As bleedwater moves upward along the form in long channels, it washes away some of the paste and exposes sand. Sand streaks can be avoided by reducing bleeding. You can do this by taking one or more of the following steps:

- Reduce water content
- Add an air-entraining admixture
- Add a water-reducing admixture
- Increase cement content
- Add fly ash to the concrete as a supplementary cementing material, or use a blended cement
- Blend a fine sand (blow sand) with the concrete sand to increase the amount of material passing the No. 50 and No. 100 screens.

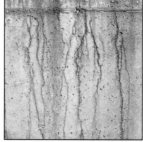

Reduce bleeding to stop streaks.

Cement scale doesn't return to zero

Q. *At our ready mix plant, even though I tare the cement weigh batcher at the beginning of each day, the scale doesn't return to zero after several loads have been batched out. The difference isn't great, but does this indicate that there's something wrong with the scale?*

A. The National Ready Mixed Concrete Association (NRMCA) *Concrete Plant Operator's Manual* says that a slow change in the cement batcher tare weight during the day is often a normal occurrence. It can be caused by material adhering to and building up in the weigh batcher. NRMCA says the condition can be temporarily corrected by re-taring the scale, but they advise a thorough cleaning to remove the accumulated material at the end or start of each day.

You can buy the *Concrete Plant Operator's Manual* for $20 (NRMCA members pay $10) plus $4.75 for shipping and handling. Send your order to Publications, NRMCA, 900 Spring St., Silver Spring, MD 20910 or fax to 301.585.4219.

Eliminating loose bolts on vibrating screens

Q. *Our operation uses a vibrating screen to classify recycled aggregate by size. We are spending a lot of time changing bolts on the screen because either the nuts come loose or the bolts elongate and become loose. What can we do to decrease downtime?*

A. The problem is caused by the extreme and prolonged vibration to which the bolts on a vibrating screen are subject. There are several solutions:
- Use hardened steel bolts. These bolts are more "stretch resistant" then standard-grade bolts. The manufacturer of the vibrating screen should be able to recommend the proper grade of bolt to use.
- Use fine-thread bolts and nuts. These can be torqued to a greater degree and are less prone to working loose. For maximum torque, tighten the nuts using a slug wrench and a sledge hammer.
- Use nuts with a nylon insert that prevents them from working loose.
- Make a short weld at the nut-bolt interface to lock the nut to the bolt. Be sure, however, to use welding rod that is compatible with the metal in the bolt and nut; otherwise the weld may crack or otherwise fail and be of no use.
- Use hydraulically applied fasteners. With these systems, a metal collar is slipped over a ribbed bolt; the collar is then "squeezed" by a hand-held hydraulic gun until it is permanently locked onto the ribs on the bolt.

A chip off the old manhole

Q. *What can be done to avoid weak top surfaces of manhole cone sections? Those surfaces tend to chip very easily. I'm enclosing a photo of one that chipped and must be repaired. We cast all of our products with a 4000 psi mix having a slump of 5 inches or less.*

Vinnie Baruch

A severely chipped manhole cover will not be accepted by contractors, and repairing the chips results in lost production time. Often the problem is solved by proper finishing after the concrete consolidates in the form.

A. Chipping usually occurs during handling or stripping of products particularly if the concrete strength is low. Even though the 28-day strength of lab-cured cylinders exceeds 4000 psi, the strength of concrete in an uncured product may be much lower than that, particularly near the top surface. After concrete has been cast, bleedwater appears at the surface, and if it does not evaporate, concrete near the top surface will have a high water-cement ratio with low strength. This condition is aggravated by troweling the top surface as soon as the concrete has been consolidated. Troweling tends to "seal" the surface and traps the bleedwater just below the surface. One or more of the following remedial measures should minimize chipping:

- If chipping occurs during stripping and handling, modify those procedures to avoid such damage.
- Use a wood or magnesium float instead of a steel trowel to smooth the surface after the concrete has been consolidated. If the surface must be troweled, wait until after bleeding has stopped, squeegee off the bleedwater, and then trowel the surface.
- Use synthetic fibers in the concrete. They help reduce chipping.

Bagging baghouse problems

Q. *Recently the plant's baghouse has not been effective in evacuating the dust at the pickup points. It doesn't seem as if there is enough air movement through our shaker-style unit. What should we do?*

A. While baghouses are relatively easy to maintain and operate, a smooth operating unit will have good air flow. High pressure drop (the difference in air pressure from the clean air intake and air exhaust) indicates that air flow is restricted. This situation may be caused by one or more of five conditions. Here's a quick checklist:

- Is the collection hopper free of buildup? Wet material caked up on the hopper sides or discharge chute restricts air flow through the exhaust. Also, clear the notary airlock, cleanout auger, and the collection hopper if necessary.
- Check the differential pressure gauge. Plugged pressure taps, leaky hoses, or a malfunctioning diaphragm in the gauge create false pressure readings. Check these items at least annually.

Griffin Environmental

Baghouses require operating adjustments depending on the moisture content of the feed material, as well as daily weather conditions.

- Check the operating adjustments of the bag-cleaning mechanism. Increasing the cleaning frequency will reduce dust buildup on the bags. Wet spring weather will require different frequency and duration settings than those in a dry summer.
- Check the age and condition of bags. Older bags might need dry-cleaning or replacement. Aging cloth loses its texture and flexibility. Over time the shaking can cause the bag fabric's weave to open up, allowing dust particles to fill in the openings.
- Is the dust falling away from the bag? Even with new or recently cleaned bags, factors can reduce a bag's surface texture and even cause the dust to stick. Be certain that the bag type is designed for the type of dust it is collecting. Most shaker-type units use bags with a polyester sateen finish that is smoother than bags used on a jet-pulse collector.
- Finally, call the baghouse manufacturer. Their lab can analyze a sample of the dust and the bag to determine dust-bag compatibility and blinding potential. Baghouses require operating adjustments depending on the moisture content of the feed material, as well as daily weather conditions.

Controlling entrained air content

Q. *We have started having problems with variations in air content of our air-entrained mix designs. Air content stays uniform for a while, then goes out of the specification range. Usually, measured air content is too low, but sometimes we get exceptionally high values. I know that many factors affect air content. Is there a systematic method for tracking down the problem?*

A. Assuming that your air content measuring methods are accurate, fluctuations in air content can be caused by changes in the materials used to make concrete, changes in mix proportions, and changes in production methods. When the dosage of air-entraining agent remains constant, expect changes in the following variables to have the effects noted.

Materials
Cement
- Air content decreases as cement fineness increases.
- Air content increases as cement alkali content increases.

Fly ash
- Air content decreases with increasing carbon content of the fly ash. Loss on ignition is a good indicator of fly ash carbon content.

Aggregate
- Air content decreases with increases in the amount of minus 200 mesh fines in the aggregate.
- Air content increases with increasing amounts of intermediate sand sizes.

Mixing water
- Air content decreases when truck mixer wash water is used as mixing water.

Other admixtures
- Air content increases when lignosulfonate water-reducers and retarders are used.
- Air content decreases when some coloring pigments are used.
- Melamine-based superplasticizers may decrease air or have little effect. Napthalene and lignosulfonate superplasticizers increase air content.
- Calcium chloride increases air content.

Mix proportions
Cement content
- Air content decreases with an increase in cement content.

Sand content
- Air content increases with increasing sand content.

Slump (Water content)
- Air content increases with increasing water content and slump up to a slump of about 6 or 7 inches. Then higher slumps decrease air content.

Production procedures

Batching sequence
- Air content decreases when air-entraining agent is simultaneously batched with cement or with other admixtures.
- Air content increases when air-entraining agent is added late in the batching sequence.

Mixer capacity
- Air content increases if the mixer is loaded to less than rated capacity and decreases if the mixer is overloaded.

Mixer speed
- Air content increases up to about 20 rpm and decreases at higher speeds.

Mixer condition
- Air content decreases if mixer blades are badly worn or if hardened mortar accumulates on the drum and mixer blades.

Temperature
- Air content decreases with an increase in concrete temperature.

Haul time
- Air content decreases during long hauls, especially during hot weather.

Don't overlook the possibility of errors in measuring air content. When the pressure method is used, an erroneously high air content will be read if the concrete is incompletely consolidated in the air meter bowl. When the volumetric method is used with extremely sticky concretes, it's hard to wash out all the air. Thus an erroneously low air content may be read. Other variations in testing method can affect the accuracy of measured air content.

The following references give additional helpful information about factors affecting air content.

References
1. S.H. Kosmatka and W.C. Panarese, *Design and Control of Concrete Mixtures*, 13th Edition, Portland Cement Association, 1994, pp. 54-59.

2. D. Whiting and D. Stark, *Control of Air Content in Concrete*, National Cooperative Highway Research Program Report No. 258 and Addendum, Transportation Research Board, Washington D.C., May 1983.

3. L.R. Roberts, "Air Content, Temperature, Unit Weight, and Yield," *Significance of Tests and Properties of Concrete and Concrete-Making Materials*, ASTM STP 169-C, 1994, pp. 65-70.

4. P. Klieger, "Air-entraining Admixtures," *Significance of Tests and Properties of Concrete and Concrete-Making Materials*, ASTM STP 169-C, 1994, pp. 484-490.

5. David A. Whiting and Mohamed A. Nagi, *Manual on Control of Air Content in Concrete*, Portland Cement Association, 1998.

Stormwater legislation raises questions

Q. *As long as the outfall at my concrete plant is clean enough to meet the effluent (partially or completely treated wastewater) limit in my permit, will I always be in compliance?*

A. Not necessarily. An outfall must meet both the effluent limit in the permit and the instream water-quality standards. It is possible to be within your effluent limit but exceed the stream water-quality standards. If this occurs, you are in violation. An example would be seepage through a dike that is not wide enough to remove all the sediment. If you can see the sediment seeping through the dike, you may be violating the instream visible turbidity standards even though you may not be in violation of an effluent limit.

Q. *We are laying out an outside area for stockpiling sand and aggregate at our plant. The storage area will consist of open stockpiles replenished by belt conveyors. Our question is this: If our maximum stockpile height is 50 feet, how do we determine the diameter of each pile at its base? Also, how do we calculate how many cubic yards and tons of material each stockpile contains?*

A. Since you already know the maximum height of the stockpiles, the next step is to calculate the diameter of the base of each pile. Do this by first determining the angle of repose of the material being stockpiled. The angle of repose is the angle between the horizontal plane of the ground and the slope of the stockpile. For aggregate, this is generally between 30 and 50 degrees. For sand it is generally between 20 and 45 degrees, depending largely on moisture content. Because the angle of repose may change based on moisture content, it may be best to determine the base diameter using the smallest angle of repose.

Knowing the height of the pile and the angle of repose allows you to determine the base diameter of each pile. You will next need the tangent of the angle of repose. The tangent of an angle is a trigonometric function and can be found in a table of natural trigonometric functions or by using a basic scientific calculator (many of which can be purchased for under $25). Following are two sample calculations:

Sand

$r = h/\tan \alpha$

r = radius of the pile

h = height of the pile

$\tan \alpha$ = tangent of the angle of repose

r = 50 feet/tangent 35 degrees

r = 50 feet/.70021 = 71.4 feet

diameter = 2 × 71.4 feet = 142.8 feet

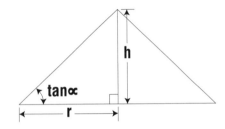

Aggregate

r = 50/tangent 40 degrees

r = 50/.83910 = 59.6 feet

diameter = 2 × 59.6 feet = 119.2 feet

To determine how many cubic yards or tons a stockpile contains, first determine the pile's volume by using the formula for the volume of a cone:

$V = \frac{1}{3}\pi r^2 h$

Continuing with the two examples above:

Sand

V = ($\frac{1}{3}$ × 3.14)(71.4^2)50 = 1.046 × 5,097.96 × 50 = 266,623 cubic feet

To determine the pile's cubic yard capacity, divide 266,623 by 27 (the number of cubic feet in a cubic yard). The result is 9,875 cubic yards.

If the sand weighs 2,750 pounds per cubic yard, the capacity of the pile = 9,875 cubic yards × 2,750 pounds per cubic yard = 27,156,250 pounds, or 13,578 tons (divide by 2,000).

Aggregate

V = ($\frac{1}{3}$ × 3.14)(59.6^2)50 = 1.046 × 3552.16 × 50 = 185,778 cubic feet, or 6,881 cubic yards.

If the aggregate weighs 2,800 pounds per cubic yard, the capacity of the pile = 6,881 cubic yards × 2,800 pounds per cubic yard = 19,266,800 pounds, or 9,633 tons.

Chapter 19

Reinforcement

Q. *The project inspector is requiring us to wire-brush mill scale and rust off all our rebar. Although the rebar has been at the site for a couple of weeks, we don't think the rust is that heavy or will interfere with the bond between the concrete and steel. Any suggestion? We're tired of brushing rebar.*

A. Fortunately, there are a couple of standards to assist you. The ASTM standard specification for deformed steel reinforcement and the Concrete Reinforcing Steel Institute (CRSI) *Manual of Standard Practice* both give the same information: Reinforcing bars with rust, mill scale, or a combination of both should be considered as satisfactory, provided the minimum dimensions, weight, and height of deformation of a hand-wire-brushed test specimen are not less than the applicable ASTM specification requirements. This inspection criteria recognizes studies that have shown mill scale and rust enhance the bond between concrete and steel.

Q. *We just placed a 6-inch-thick slab on grade with #4 rebars spaced at 12 inches on centers each way. The rebars were 1¼ inches from the slab surface. The concrete was placed by crane and bucket and had a 5- to 6-inch slump. It was hand-screeded without vibration.*

After the slab had hardened, visible lines marked the location of the rebars and the owner is upset. What caused the problem? Is there a way of removing the lines?

A. What you described is sometimes called steel shadowing. There are several theories as to how it happens. Some believe that rebar heats up when the sun hits it. This then causes the concrete to hydrate more rapidly over the rebar. When lines have shown on the underside of elevated slabs it's been blamed on form oil being chemically changed by the sun but unaffected in areas shaded by the rebar. In one case the shadow of a telephone pole on a flatwork job left a long visible streak in the concrete, seeming to confirm this possibility.

Another theory holds that when rebar is too close to the surface and the concrete isn't properly vibrated, the concrete doesn't fully knit together over the bars. This causes the visible line.

Like most discolorations that aren't stains, these marks are probably not erasable. You can cover them with an opaque coating, but that's a pretty expensive solution for what is essentially a cosmetic problem.

Reader response: Reader Mike Jacobson with R.P. Carbone Construction Co., Cleveland, submitted the following comment on the November 1992 Problem Clinic item, "Rebar Near Surface Causes Marks in Slab" (page 846):

During concrete placement and consolidation, larger aggregate particles will be displaced from the area directly over the bars. The concrete mix becomes nonuniform and hydrates at different rates, resulting in a different color, or shadowing. Shadowing can be minimized by maintaining rebar alignment and coordinating cover and maximum aggregate size.

Some rust on rebar is acceptable

Rebar near surface causes marks in slab

Dowels or keys in a 4-inch slab?

Q. *We are placing 4-inch-thick slabs. Can we use dowels or keyways for load transfer?*

A. In ACI 302.1R-89, "Guide for Concrete Floor and Slab Construction," ACI Committee 302 does not recommend keyed joints for slabs thinner than 6 inches, and also doesn't provide guidance for dowels in slabs thinner than 5 inches. We've seen metal key forming systems with knockouts for dowels advertised for 3- and 4-inch slabs, but we don't have information on field performance of these joints. You might consider thickening your slab edges, beginning 2 or 3 feet from the joint, and detailing a doweled joint in a 5-inch slab.

Cutting dowel bars

Q. *Is it OK to shear steel rods when fabricating dowel bars for concrete slabs?*

A. Unless there are further finishing operations for the dowel, avoid sheared dowels. Shearing tends to form a lip at the end of the dowel. This will cause the bar to hang up when the slabs move apart and may cause a transverse crack along the dowel ends. Smooth dowels are intended to make adjacent slabs or pavements share loads by deflecting together. There should be little restraint along the length of the dowel.

Old reinforcing bar systems

Q. *Where can we find information on systems for reinforcing building slabs built in the 1930s, and earlier?*

A. One reference is the Concrete Reinforcing Steel Institute (CRSI) publication, *Evaluation of Reinforcing Steel Systems in Old Reinforced Concrete Structures*. The 16-page publication can be ordered from CRSI, 933 N. Plum Grove Rd., Schaumburg, IL 60173 (847-517-1200; fax 847.517.1206).

In evaluating these old systems, you may find that the bar layout and mechanical properties tensile strength and bond are significantly different than those used in present practice.

What's Type N reinforcing steel?

Q. *While inspecting reinforcing bars that were to be used on a project, I found several bundles designated N for type of steel. I know that the designations S, I, A and W indicate billet, rail, axle and low-alloy steels, respectively. But I can't find any ASTM reference to Type N. What kind of steel is it?*

A. Anthony Felder of the Concrete Reinforcing Steel Institute says that the bars are made of billet steel. The N stands for new billet steel. Until 1984, ASTM A 615, "Standard Specification for Deformed and Plain Billet-Steel Bars for Concrete Reinforcement," included two types of billet steel, N and S. Type S steel met more stringent supplementary requirements, which were optional. In 1984, these supplementary requirements were moved into the main body of the specification, the Type S designation was kept for consistency, and the Type N designation was dropped.

You can conclude, therefore, that the steel you inspected was produced in 1984 at the latest, and it may not meet the current ASTM A 615 requirements.

Q. *From time to time I see references to "half cell" readings as a measure of corrosion activity in concrete. Often there is computerized equipment for recording and analyzing the data, and it looks like a "black box" to me. How does this system really work?*

A. To understand these half cell devices it's important to remember that the corrosion of reinforcing steel in concrete is an electrochemical action. Both chemical processes and a flow of electricity are involved. The difference in electrical potential at various points on the rebar generates a flow of current from one point to another, forming an electric cell, also called a galvanic cell.

Detecting and measuring this current flow helps investigators assess the degree of unseen corrosion. Locations for readings are marked on the concrete, usually in a grid pattern. Then the negative terminal of a direct current voltmeter is electrically connected to the reinforcing steel mat. The operator moves a copper-copper sulfate half cell connected to the positive terminal of the voltmeter from point to point on the grid. The voltmeter readings at the grid points are recorded in millivolts. For readings of 200 to 300 millivolts or higher, depending on the type of structure, investigators are concerned about hidden corrosion activity and conduct other tests to confirm the diagnosis. ASTM C 876, "Standard Test Method for Half-Cell Potentials of Reinforcing Steel in Concrete," sets rules for testing and indicates probability of corrosion for different levels of voltage readings.

Q. *We're going to be placing a 6-inch-thick concrete slab on grade that requires welded wire fabric for crack control. Plans call for it to be in the center of the slab. We think it ought to be closer to the top. Also, we want to place the concrete in two layers, laying mats of the fabric between them. But the architect says we can't do it that way because we may get a cold joint where the two layers meet. Is he right?*

A. The Wire Reinforcing Institute recommends locating welded wire fabric 2 inches below the surface in a 6-inch-thick slab. They also say that the two-course method is probably the most effective way of placing fabric, although it takes more time. Place the top course before the lower concrete starts to set to avoid producing a cold joint.

Q. *We're building a floor for which the plans call for keyed construction joints. The plans also call for wire mesh that runs continuous through the joint. We don't think this is a good detail either for constructibility or for floor performance. How can we convince the engineer that there are better ways to do the jointing work?*

A. Keys are not effective load transfer devices at construction joints if the joints open up too much when shrinkage occurs. That's probably why the designer wants to run wire mesh through the joint to prevent it from opening too wide. But if the joint also is to serve as a control joint, the wire mesh defeats the purpose of the joint. It restricts joint movement and may cause random cracks at other locations.

The detail also is expensive to build because you'll have to split the wood bulkhead used to form the keyway to let the wire mesh pass through it. That will increase the chances that the floor won't be as flat as it could be at the bulkhead.

Here are two recommended ways of handling the construction joint that are better than a keyed joint with wire mesh running through it:

- If the construction joint *does not* have to serve as a control joint, use deformed steel bars to tie the two joint faces together and ensure shear transfer. If the engineer wants a keyed joint with steel, drill holes in the keyed bulkhead and run short deformed steel bars through the holes. Then run the wire mesh up to the bulkhead and start it up again on the other side.
- If the construction joint *does* have to serve as a control joint (permitting movement), then pass smooth steel dowels through the bulkhead, but don't run the mesh through the joint. If you use smooth steel dowels that are greased on one end, you don't need a keyway.

Can you tack weld rebar cages?

Q. *What's the current thinking on effects of tack welding on rebars? I've heard for years that it weakens rebars, but I still see it used in making rebar cages.*

A. Section 7.5.4 of "Building Code Requirements for Structural Concrete" (ACI 318-95) says welding of crossing bars shall not be permitted for assembly of reinforcement unless authorized by the engineer. Tack welding can seriously weaken a bar at the point welded by creating a metallurgical notch effect (see Figure) according to the code commentary.

Article 5.4 in the American Welding Society's welding code for reinforcing bars (AWS D1.4) also prohibits tack welds that don't become a part of permanent welds, unless authorized by the engineer. However, because tack welding can be authorized by the engineer, there are instances where it is acceptable. Usually approval will be given for tack welds at points of low design stress in each crossing bar or where ductility of the bars isn't especially important. AWS D1.4-92 states that tack welds shall be made using preheat and welded with electrodes meeting the requirements of final welds. They shall be thoroughly cleaned and subjected to the same quality requirements as the final welds.

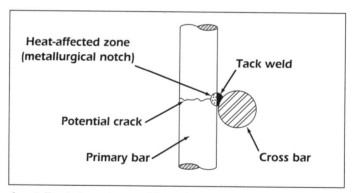

A metallurgical notch caused by tack welding may cause brittle failure of rebar cages under accidental impact.

Fabrication practices in Europe include extensive use of tack welds. But European standards for reinforcement typically include controls on chemical composition of the bars. Chemical composition and carbon equivalent have an effect on weldability. In the United States, ASTM Specification A 706 for low-alloy steel reinforcing bars includes limits on chemical composition and carbon equivalent.

Billet-steel bars (ASTM A 615) should be welded with caution because the A 615 specification includes no provisions to enhance weldability. The specification says that when this steel is to be welded, a welding procedure suitable for the chemical composition and intended use or service should be used. AWS D1.4-92 describes the proper selection of filler metals, preheat temperatures, performance requirements, and requirements for procedure qualification. You can order a copy of AWS D1.4 from American Welding Society, 550 N.W. LeJeune Rd., Miami, FL 33126.

References

1. *Building Code Requirements for Structural Concrete*, ACI 318-95, American Concrete Institute, 1995.

2. *Structural Welding Code Reinforcing Steel* (ANSI/AWS D1.4-92), American Welding Society, 1992. A metallurgical notch cause by tack welding may cause brittle failure of rebar cages under accidental impact.

Q. *A contractor in our area places wire mesh reinforcement in residential slabs on grade ½ to 1 inch from the surface of the concrete. The contractor justifies this practice by saying that if the wire mesh isn't high enough within the slab it will not be able to keep cracks closed tightly. We think he is placing the wire mesh too high. Is placing wire mesh ½ to 1 inch from the surface of a residential slab on grade acceptable practice?*

A. If wire mesh is located too close to the surface of a slab it may corrode, especially in exterior applications where the concrete isn't continually dry. ACI 332R-84 "Guide to Residential Cast-in-place Concrete Construction" states that for lightly reinforced, 4-inch-thick slabs (Slab Type B) reinforcement should be placed in the middle of the slab, a minimum of 2 inches from the top surface.

Where should wire mesh be located?

Q. *I know that concrete sets due to a chemical reaction between the cement and water. But does water-cement ratio also affect the setting time?*

A. Setting of concrete involves both chemical and physical phenomena. Several studies have shown that when the water-cement ratio is reduced, the initial and final times of setting are also reduced.

In one study, water-cement ratio was increased by increasing water content while keeping all other batch quantities constant (Ref. 1). The effect on time to reach initial and final set is shown in the following table:

Does water content affect setting time?

Water-Cement Ratio	Setting Time	
	Initial	Final
0.53	4 hr. 25 min.	7 hr. 15 min.
0.57	4 hr. 55 min.	7 hr. 45 min.
0.61	5 hr. 20 min.	8 hr. 10 min.
0.67	5 hr. 40 min.	8 hr. 30 min.

In this study, concrete temperature was 63° F, ambient temperature was 70° F, and ambient relative humidity was 70%.

Another study showed similar results (Ref. 2). Concretes made with 517 pounds of Type I cement per cubic yard and water-cement ratios of 0.54 and 0.58 had initial setting times of 4 hours 40 minutes and 5 hours 10 minutes respectively. Final setting times were 6 hours 20 minutes and 6 hours 45 minutes. The tests were run at 73° F.

Results of these studies support the rule of thumb that adding a gallon of water per cubic yard of concrete can be expected to increase setting time by about 15 minutes at moderate temperatures.

References

1. G.D. Stefanou and Ch. Larsinos, "Influence of Mixing Water on the Setting Time of Concrete," *The International Journal of Cement Composites and Lightweight Concrete*, February 1981, pp. 45-48.

2. Vance Dodson, "Time of Setting," *Significance of Tests and Properties of Concrete and Concrete-Making Materials*, STP 169C, ASTM, West Conshohocken, Pa., 1994, p. 83.

Cause for slow setting

Q. *I'm a pattern-stamping contractor in Louisiana and last winter I had a problem with slow setting and excessive bleeding of a concrete slab. The weather was unusually cold here, with air temperatures around 45° F. The cementitious material was 20% fly ash, and the concrete also contained a water reducer. The concrete didn't set for six hours, and there was so much bleedwater that dry color-shake application and stamping were delayed. What caused this to happen?*

A. The three factors that you mentioned—low temperatures, replacing cement with fly ash, and use of a water reducer—may all have contributed to slow setting of the concrete.

When concrete sets slowly, the bleeding period is prolonged. That's why you had more bleedwater than usual. To avoid this problem during cool weather, you can order concrete that doesn't contain a water reducer, or have the producer add an additional 100 pounds of cement per cubic yard of concrete. Don't use calcium chloride accelerators because they can cause discoloration of decorative concrete flatwork.

Chapter 20

Safety

Q. *Safety inspectors told us that the Occupational Safety and Health Administration now prohibits the use of plastic mushroom caps for protecting workers from impalement on projecting rebar. They said troughs that cover a whole row of projecting rebar will meet OSHA's test criteria of a 250-pound weight dropped from a distance of 10 feet without rebar breakthrough. Where can we purchase these troughs?*

A. We know of four sources for plastic models of this device, and there may be others. The manufacturers' literature should tell you whether the models meet OSHA test criteria.

Aztec Equipment Inc.
P.O. Box 820
Bloomington, CA 92316
(800-531-3355)

Norseman Plastics
39 Westmore Dr.
Rexdale ON M9V 3Y6 Canada
(800-267-4391)

Deslauriers Inc.
P.O. Box 189
Bellwood, IL 60104
(800-743-4106)

Paragon Products
1516 E. Mapleleaf Dr.
Mount Pleasant, IA 52641
(800-776-0385)

We couldn't find a commercial source for wooden troughs, but did learn that you can make your own by nailing together three 2x4s or by routing a continuous slot in the edge of a 2x4. Check with your OSHA compliance officer to make sure any homemade wooden troughs meet OSHA requirements.

Q. *How frequently should I replace my crew's hard hats?*

A. The OSHA part 1926 regulations don't specify a service life for hard hats. Frequent inspection of the hard hat for cracks, serious chips, or wear, and replacement of damaged hats is a good approach. Why not make hard hat checks part of your regular tool box safety meetings?

Q. *I'm worried about a disgruntled former employee anonymously filing a bogus complaint against me with OSHA. If he makes a non-formal complaint, will OSHA make a full-blown inspection, or will the enforcement officer inspect only the complaint?*

Sources for rebar impalement-protection devices

Replacing hard hats

Do complaints trigger inspections?

A. You may be able to avoid an inspection. If a non-formal complaint is made against you, except in imminent danger situations, OSHA will first send you a letter asking you to investigate the claim and correct the problem. You will need to write to OSHA, telling them what you have done to correct the problem or proving there is no problem. If possible, photographs should accompany the letter. While this does not guarantee that you will avoid inspection, it greatly reduces the possibility. If the OSHA office believes the problem has been corrected, only one in 10 plants will be chosen for inspection at random. If you are inspected, there are several items that OSHA always checks in addition to the items on the complaint:

- Hazard Communication program (HazCom)
- Lockout/tagout program
- Your record keeping and lost-time injury rate

If an enforcement officer does come for an inspection after a non-formal complaint, OSHA may cite you for violations in plain view.

Maintenance of precast strand chucks

Q. *What should we do to maintain our prestressed strand chucks in proper working order? These pieces of equipment hold thousands of pounds of force, while people work nearby unaware of the danger if one of these strands breaks. Our company doesn't have a maintenance program for strand chucks and I would like to develop one.*

A. You are right in saying strand chucks are precision pieces of equipment and often they are not given the respect they deserve. The Precast/Prestressed Concrete Institute gives the following procedure on inspecting and lubricating the chucks:

1. Separate all chucks by size. Inspect one size of chuck completely before placing another size on the bench.

2. Disassemble all chucks by removing the jaws and the rubber retaining ring.

3. Put on safety goggles before using the brushes to clean the chucks. Rubber gloves that will not get caught in the brushes should also be used.

4. The chuck body should be cleaned with a tapered wire brush.

5. Before removing the rubber retaining ring, the jaws can be cleaned with a motorized tapered nylon brush. Care should be taken to assure the wire body brush is never used to clean the jaws as damage could result.

6. After cleaning, remove the rubber retaining ring and inspect the jaws for signs and scratches, chipped threads, score marks. Examine the rubber retaining rings for signs of splitting or damage. Inspect the barrels for signs of excessive rust, corrosion, or dents. Discard all damaged parts to assure they are not used in reassembly.

7. Clean and lubricate the barrels and allow them to dry approximately five minutes. Lubricate the exterior of the jaw assembly.

8. Reassemble the chucks and store them in a rack designed to keep them clean and out of the weather.

In addition, Rex Hartup of Prestress Supply Inc., a supplier of strand chucks and related equipment, suggests that chucks be cleaned, inspected, and lubricated after each use. Proper care and maintenance will extend service life. Cleaning removes contamination that interferes with the jaws gripping the strand and correct seating of the body. Chucks must be disassembled, cleaned, and inspected after each use, especially when using spray lubricant or graphite. The units must be thoroughly cleaned before applying more lubricant to avoid excessive buildup. After cleaning, store the chucks in diesel oil if you don't plan to use them for more than one week. This will keep the jaw teeth from rusting and reduces slippage. If worn-out parts are found during inspection, they should be replaced. Lubrication allows proper seating and reduces the danger of slippage and damage to the jaw assemblies.

Reference

Manual for Quality Control for Plants and Production of Precast and Prestressed Concrete Products, Precast/Prestressed Concrete Institute, Publication Number MNL-116-85.

Q. *When I look at the mill test report for my Type I portland cement, the chemical composition section listed a value of 21% silicon dioxide. Does this mean there are dangerous levels of respirable silica in the cement?*

A. No. The oxide analysis in a mill report tells what percentages of elements, expressed as oxides are in the cement. Limestone, clay and sand are commonly used in making portland cement. In their raw form, these materials contain varying amounts of silicon dioxide. But when they're blended and heated to high temperatures in the cement kiln, the resulting clinker contains new compounds. What was originally crystalline silica has reacted to become part of the hydraulic calcium silicates.

According to Howard Kanare, group manager of chemical services at Construction Technology Laboratory (subsidiary of the Portland Cement Association), Skokie, Ill., most ASTM C 150 Type I portland cements contain less than 0.1 % crystalline silica by weight. "Gypsum is added to the clinker before final grinding. We believe impurities in the gypsum are the source of the very small amount of crystalline silica," says Kanare.

Producers should review recently revised Material Safety Data Sheets from their cement suppliers. OSHA has required manufacturers to clearly list any potential hazard from crystalline silica.

Q. *We plan to take several cores from a building slab and want to avoid making bolt holes in the concrete surface to secure the drill stands. I don't want my workers standing on the drills to keep them in place. Are there other ways to secure the drills?*

A. Most drills can be mounted on a vacuum base having a rubber gasket beneath it to hold the drill in place. A second approach is to wedge a piece of conduit or pipe between the drill stand and the ceiling. Many core drilling machines have a ceiling jack screw for this purpose.

Crystalline silica and portland cement

Securing drill stands without bolts

Chapter 21

Stain Removal/Cleaning

Q. *We're making precast panels using integral color in the concrete. The panels are covered with a tarp at night to hold in the heat and make them cure faster. Our problem is that moisture condenses on the underside of the tarp and drips on the panel. When the drops evaporate they leave a white powder that stains the panel surface. What can we do to prevent the stains or to remove them afterwards?*

Stains in precast panels

A. The stain is probably an alkaline salt that has leached out of the concrete. When water evaporates, it doesn't carry the salts with it. So I'd guess that drops of pure water on the surface cause salts from within the concrete to migrate toward the pure water. Some cements cause greater amounts of efflorescence than others, so you might solve the problem by changing cement source. Try a low-alkali cement if one is available. Changing cements, however, isn't a good idea if you've already cast a number of panels. The new ones probably won't match the color of the ones already cast.

Another option is removing the efflorescence by washing and brushing with a stiff-bristle brush. Do this as soon as possible after form stripping when the salts are possibly still water soluble. An acid may work if plain water doesn't dissolve the salts. Try diluted vinegar first. If that doesn't work, a muriatic acid may be needed, followed by a neutralizing solution. All of these cleaning methods are labor intensive and will drive up your cost per panel. Also, it's important to standardize the procedure so washing itself doesn't cause color changes.

A third option is using the panels as is. Discuss the situation with your client. Stains of this type may become less noticeable with the passage of time. If all panels look nearly the same, the condition may be tolerable.

Reader response: I found the answer to the question on stains in precast panels in the July 1992 Problem Clinic (*Concrete Construction*, page 564) inadequate. A study conducted in the early 1960s, which I started and which was completed by N. Greening and R. Landgren, covered this specific problem, and others, very well. The report, "Surface Discoloration of Concrete Flatwork," is published by the Portland Cement Association (Research Dept. Bulletin 203).

The authors call this type of discoloration the "greenhouse effect." In their words: "A high fold in a waterproof curing sheet can serve as a little 'greenhouse.' On a hot day, under direct sunlight, the fold becomes the location of a water evaporation-condensation cycle. The heat of the sun, aided perhaps by the heat of hydration of the concrete, evaporates water from the concrete under the fold. The water vapor then condenses on the cool high part of the fold and eventually runs down the sides of the film to collect at the points of intersection of the concrete and the film, or in low places in the concrete surface. Such localized dry and wet areas on fresh concrete surfaces may cause concrete discoloration."

The authors were able to create such discoloration in the laboratory with the use of a polyethylene sheet, a sun lamp, and a fan. They observed the cycling of water evaporating from the concrete in the tented or "greenhouse" areas, condensing on the plastic and draining down to become part of the water film between the concrete and those portions of the plastic that were lying flat. Continuing, the authors stated: "Dark areas occur wherever the polyethylene remained sufficiently out of contact with the concrete surface to permit evaporation to occur. Subsequent scrubbing with water, using a stiff bristle brush removed some of the concentrated white efflorescence deposits. Application of very dilute hydrochloric acid removed the remaining efflorescence."

The material removed by the water was considered to be sodium and potassium salts (hydroxides and carbonates) and the material removed by the acid wash was considered to be calcium salts.

Calcium chloride in the mix was found to cause a dark mottling beneath the efflorescence that was difficult to remove.

I also was disappointed in your advice to the reader on how to avoid the discoloration. The obvious answer is to avoid allowing the plastic sheet to contact the concrete surface sporadically. This is difficult. An easier solution is to use a layer of burlap between the concrete and the plastic or, if necessary, switch to water curing or a curing compound.

William F. Perenchio, Wiss, Janney, Elstner Associates Inc., Northbrook, Ill.

Stains from straw

Q. *We read your recent article on cleaning concrete (Concrete Construction, November 1992, page 791) thoroughly but failed to find any recommendations on the removal of straw stains from concrete. In the past year we've had several requests for help with yellowish brown stains caused by straw, but we haven't found a good remedy.*

A. We've examined all our information sources on stain removal, some from Europe as well as the United States, and find no reference to stains caused by straw. Scrubbing with a strong detergent solution might remove the stain, particularly if the stain is fresh. This is one of the removal methods least harmful to the concrete surface.

Removing keel marks from formed surfaces

Q. *On an elevated deck job we're doing, ironworkers used black lumber crayons on the deck forms to mark the location of rebar. When we stripped the forms we found that the black marks had transferred to the concrete surface on the underside of the deck. They are very noticeable and unattractive. How can they be removed?*

A. We suggested using a commercial solvent but that didn't do the job. The contractor then found his own solution. He used a 3000-psi water wand and put a mild cleaner in the water. That removed the marks.

Removing excess bond breaker from tilt-up panels

Q. *I saw your Problem Clinic regarding too much bond breaker being used on tilt-up panels (Concrete Construction, January 1996, pp. 76-82). We have this problem on a job. Do you know how to get the bond breaker off the panels so they can be successfully painted?*

A. Many manufacturers of curing compounds, sealers, and bond breakers also make strippers that can remove their products. We suggest asking the bond-breaker manufacturer for advice on the best chemical stripper and cleaning method. Stripper effectiveness varies with bond-breaker composition. Chemicals that have been used in strippers include D-limonene (a citrus-peel-based stripper) and methylene chloride.

Q. *I run a pressure-washing business in Georgia. To remove the red clay that stains many driveways, I wet the driveway, then use a low-pressure hand sprayer to apply a dilute solution of muriatic acid (one part commercial-grade muriatic acid to three parts water). After that, I immediately rinse the concrete surface with a pressure washer. I can achieve up to 3000-psi pressures, but seldom have to approach that to get the driveway clean. I've washed hundreds of concrete driveways using this method and never had a problem.*

Last June, however, I pressure washed a one-year-old driveway, and five months later the homeowner said it had ruined the concrete surface. I inspected the concrete and did find that the surface is eroding in spots. The erosion is ⅛ to ³⁄₁₆ inch deep, exposing coarse aggregate. Since I've never had this problem before, I think the concrete must have been of poor quality. Is there any other explanation?

A. We asked consultant Robert Cain, KRC Associates, Milford, Ohio, about the possibility that residual acid had eroded the concrete surface to the depth you describe. He says that's highly unlikely because pressure washing would have removed any acid that wasn't neutralized by reaction with the concrete. However, Cain says acid opens up the concrete surface and advises that you apply a sealer after the concrete dries. This will help keep the driveway clean for a longer period between washings.

We recommend that you hire a materials consultant to examine the distressed portions of the driveway and, if necessary, take cores that could be used to determine concrete quality.

Q. *As a materials engineer, I am trying to discover the cause of discoloration. The contractor involved is very experienced in tilt-up construction and reports that this discoloration has only occurred three or four times over recent years. The concrete supplier operates a very good quality-control program. The concrete used for the panels does not incorporate superplasticizer, and no admixtures were added onsite. The concrete was also well-mixed and placed at a slump of 4 to 5 inches.*

The discoloration shown is evident only on the underside of the tilt-up panels (the side cast against the slab). The dark patches are characterized by a surface full of pinholes and a weak matrix, which is easily abraded by rubbing with a finger. The dark patches are more common around the perimeter of the tilt-up panels. They also seem to occur at high spots on the panels (which correspond to low spots in the slab), but there are some exceptions. The panels were cast directly on the slab, which was coated with a bond breaker. The contractor indicates that the concrete was thoroughly vibrated with internal vibrators, including the use of a poker vibrator along the edge forms.

I would like to know if any readers have experienced this phenomenon, especially if they have learned the cause.

A. Excessive release agent was applied to the slab, and it ponded in the low spots. Too much release agent can act as a retarder. The solution is to make certain the release agent is applied according to the manufacturer's recommendations and that excessive amounts of release agent are not allowed to accumulate.

James M. Shilstone Jr., The Shilstone Companies Inc., Dallas

The cause is too much bond breaker. The reason the dark patches occur around the [panel] perimeter and at low spots on the slab is because bond breaker was applied so heavily that it puddled. The solution is to instruct workers to use less bond breaker and to spread it out thinly and evenly with a household mop or rags.

If the pinholes and a weak matrix persist, the bond breaker has too high a fatty-acid content; get a higher-quality bond breaker.

David G. Markle, Tulsa Dynaspan Inc., Broken Arrow, Okla.

Can acid cleaning cause driveway deterioration?

Readers' solutions to "What caused dark patches on tilt-up panels?"

Improper bond-breaker application and the osmotic effect are possible causes of the patches and pinholes on this tilt-up panel.

The problem is not in the concrete mix. If it were, the discoloration would be consistent throughout the thickness of the panel. I believe the problem is an excessive amount of bond breaker, which can tend to act as a retarder. As the panel forms were sprayed, the excess oozed down to the slab. As the slab was sprayed, the excess shed from high points of the slab to low points. Solution: Pour a flatter slab (to eliminate low points), and spray it with two light coats of bond breaker prior to rebar placement. With bond breaker, more is definitely not better.
Ted N. Mefelli, La Russo Concrete Co. Inc., Lakewood, Colo.

It appears that the panels were either coated with an inappropriate bond breaker, or an overabundance of bond breaker was used and [excess] was not sufficiently wiped off prior to casting the concrete. Bond breaker, if not used properly, can act as a set retarder, causing powdering of the concrete surface in areas where there may have been a buildup. The solution is to use a drying type of product, spray it on lightly (more is not better), and wipe off all puddles or overabundance of bond breaker prior to concrete placement.
William Pompili, Pompili Precast Concrete, Garfield Heights, Ohio

If the contractor applied one heavy coat of the bond breaker and it settled in the low spots on the slab, it may easily have had a retarding and mottling effect on the panels. Bond breaker should be applied to the slab in two light coats in opposite directions. Application of a single heavy coat, especially if done in cool and damp weather, may have resulted in ponding and uncured bond breaker left on the slab when the panels were poured. To test bond breaker [application], splash a little water on the slab. If a bond breaker has been properly applied, the water tends to bead up like it does on a freshly waxed car.
William J. Zens, Allied Building Products, Seattle

Discoloration problems have been around for some time, but they have resurfaced at a more frequent rate here in Las Vegas due to more tilt-up [buildings] being placed. The Tilt-Up Concrete Association's newsletter Tilt Tips (No. 1) addresses this problem. It is called the "osmotic effect."

TCA says that surface defects found on the downside of tilt-up panels, such as spalling, delamination, or dusting, are often the result of incomplete hydration of the cement due to insufficient water. Water that would normally be available for hydration is drawn out of the panel, down into the casting or floor slab by osmosis. The result: a weakened downside surface. The newsletter also suggests an effective solution to preventing water penetration. Create a water-impermeable barrier between the casting slab and the tilt-up panel by using a membrane-forming combination curing compound and bond breaker. Apply the product according to the manufacturer's instructions as the curing agent for the casting slab and as the bond breaker prior to casting of the tilt-up panel.

I am in the process of placing some test panels using various mix designs, admixtures, and bond breakers, including a test panel half with and half without a membrane-forming curing compound.

Wayne Stroud, WMK Materials, Las Vegas

Cleaning concrete splatter off of aluminum

Q. *Our placing crew splattered concrete on an aluminum storm door when we were placing a patio. The splatters are several days old. I heard that concrete etches aluminum and need to know how to clean off the splatters without marring the aluminum surface.*

A. We called an aluminum producer and they recommended using Soft Scrub, a bathroom cleaning product you can buy at a grocery store. They said this product will shine the aluminum and can be used to treat weathered aluminum.

To test their suggestion, we put a concrete splatter on an aluminum frame (photo A), then scraped it off with a metal hand scraper four days later (photo B). A cement film still remained after scraping, and we tried to remove it with a sponge soaked with Soft Scrub. When that had no effect, we applied Soft Scrub directly on the cement film and waited a minute before wiping it off with a sponge. Neither of these methods removed the cement film (photo C).

We next used extra-fine (#000) steel wool and succeeded in removing some of the cement paste. But the underlying aluminum was darker in color, even after we scrubbed it again with Soft Scrub (photo D). However, the Soft Scrub did shine up the aluminum that hadn't come into contact with the splatter.

Removing the splatter sooner might have avoided the etching problem, but we didn't test the recommended procedure at an earlier stage.

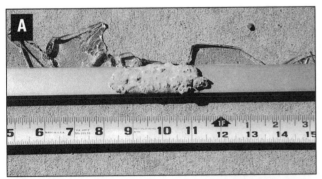

Attempts to remove concrete splatter from an aluminum frame.

Cleaning white concrete

Q. *We need to remove hardened construction adhesive from a new white concrete surface. We've tried grinding, but the material melts. Any other suggestions?*

A. Bob Joyce of Quality Restorations, Wood Dale, Ill., suggests applying dry ice to the material, which should then become brittle and easier to remove. As with the removal of any stain, practice your procedure on a small area first.

Removing rubber tire marks

Q. *Do you have any suggestions for removing rubber tire marks from precast concrete pavers?*

A. We asked Norm Gill from Tennant Co. to answer your question. He recommends using a citrus-based alkaline cleaner (Tennant 9960 Severe Soilage Neutral Floor Cleaner is one such product), and says to spray the concentrated cleaner on the surface and allow it to soak the tire marks for 15 minutes. Use a mechanical floor scrubber to remove the marks.

Solvents may also remove the marks, but many of these products are considered to be hazardous materials and they will leach into the soil at paver joints. Disposal of excess solvent also is a problem. Citrus alkaline cleaners aren't hazardous and, in most locations, they can safely be washed into a sewer.

Removing wood stains from concrete

Q. *After one of our customer-contractors poured an integrally colored patio at a new residential home, the homeowner had the carpenters add a railing to the adjacent redwood deck. Afterward, wood debris covered a portion of the patio. During the final walk-through at loan closing, the homeowner noticed a red stain on the tan-colored concrete. How can the contractor remove the stain to satisfy the owner without replacing the concrete?*

A. The stain in this example was caused by wood extracts. These natural oils permeated concrete surface voids, leaving the brown or red stains. The best way to attack the stain is with early treatment that draws the stain up from the concrete surface with a poultice.

A poultice is usually made by mixing some essentially inert fine powder, such as ground limestone, hydrated lime, portland cement or fuller's earth, with the solvent or solution. (Shredded pieces of highly absorbent paper would also work.) The selection of the cleaning liquid is based upon the cause of the stain. The two parts are blended to make a smooth paste that is troweled over the stain. The liquid portion of the poultice migrates into the concrete, where it dissolves some staining material. Then the liquid gradually retreats upward from the concrete into the poultice, from which it evaporates, leaving its burden of dissolved staining material in the poultice's inert powder. Once dry, the residue is scraped or brushed away. Usually it takes more than one application to remove all the stain. Whether the concrete surface is colored or not, it is important to first test the poultice in an out-of-the-way small area to check its effect on the concrete's surface. To remove the wood extracts' residue, first use a soft brush to wash the stained area thoroughly with a solution that is 1 part glycerol in 4 parts water. Treat the stained area with a poultice made with a diluted hypochlorite solution, made up of household bleach (which is usually about 5% sodium hypochlorite) and tap water, mixed at a ratio of 1 part bleach with 4 to 6 parts water.

Reference
"Methods of removing some specific stains from concrete: Oil to Wood," W. Kuenning, *Concrete Construction*, September 1986, pp 821-826.

Q. *We would like to remove some unsightly stains from a precast concrete building. What is the most effective way to remove these stains?*

A. Green and sometimes brown stains on concrete are common where water has flowed over copper or bronze. These stains can be removed by using a poultice made by dry mixing one part by weight of ammonium chloride or aluminum chloride with four parts by weight of fine-powdered inert material, such as diatomaceous earth or talc.

Form a smooth paste by combining the previous two ingredients with ammonium hydroxide (one part concentrated ammonium hydroxide diluted with two to nine parts water). If the concentrated solution is not available, use household ammonia without diluting. Apply the poultice over the stain to a thickness of ⅛ to ¼ inch and leave until dry. Then scrub the surface with scouring powder and water. Three applications should remove even severe stains. If the needed chemicals are not available at your local hardware or drug store, they can be ordered through a chemical or pharmaceutical supply house.

Removing copper stains from precast concrete

Q. *We recently placed and finished a concrete sidewalk. After curing the concrete for one day, we covered it with plywood to protect the surface. It rained that evening and when we removed the plywood the following day, the concrete had brown stains on it. How can we remove the stains?*

A. Extracts from wood may cause yellow or brown stains in concrete. The stains can be removed by treating the concrete with a hypochlorite solution. Household bleach, which is usually about 5 percent sodium hypochlorite, is satisfactory after diluting one part of bleach with four to six parts of water. The stain disappears if the surface is first scrubbed thoroughly with a solution that is one part glycerol in four parts water. Glycerol is flammable and bleach is corrosive to skin. As always, read the warning labels on the chemicals before use.

Removing plywood stains from sidewalk slabs

Q. *What are some alternate ways for removing oil stains from a concrete floor? We've had pretty good results from scrubbing with trisodium phosphate (TSP) and hot water but the method is time consuming and laborious.*

A. If it's a relatively fresh oil spot and won't be exposed to moisture, try dusting on portland cement. Leave the dry cement on the stain overnight then sweep it off and pick it up with a dust pan. Repeat the process if all the oil isn't removed. This method reportedly draws oil out of the concrete.

Removing oil stains

Q. *We recently removed stacks of architectural precast sound wall panels that were resting on freshly cut oak timbers. The bottom panels all have a sticky, glue-like substance on their surface. Removing the substance by sanding or grinding will ruin the architectural finish. In addition, the panels are colored with an integral red oxide. Is there a chemical we can use to remove this substance without ruining the appearance of the panels?*

A. According to Andy Baker, a chemical engineer at the United States Forest Products Laboratory, Madison, Wis., the substance on your precast panels is pitch, a resinous material excreted by trees to seal their damaged areas. Pitch will dissolve in both gasoline and methylene

Removing wood pitch from architectural precast panels

chloride. If trying gasoline, use a "clean" gasoline like the white gas used in camp stoves. This will leave less of a residue on the panels after evaporation. Clean the panels in an open, well-ventilated area, with no sources of sparks or open flames.

Methylene chloride is an active ingredient in paint remover and can be found at most scientific supply stores. It is not flammable, but exposure to the fumes can lead to kidney disease or heart problems. Wear neoprene gloves when using it. Wear a respirator when using it indoors and read the warning label carefully.

Gasoline and methylene chloride may discolor concrete. Use them first on an inconspicuous location to see the effect on the panels' appearance.

Repairing a discolored driveway

Q. *The photos show a driveway with several very light spots that look as if the concrete has been bleached. The driveway is 2 years old and the spots are as bad now as they were right after it was built. The owner is upset and wants the driveway replaced but it's crack free and performing perfectly except for this cosmetic problem. The owner says if he scratches the gray areas they're light underneath. Would grinding remove the stain? If not, do you have any other suggestions?*

A. The problem looks more like discoloration than staining. It wouldn't hurt to try lightly grinding an area adjacent to the white areas so you can see if that helps. If that doesn't work, a traffic-resistant coating will cover the whole driveway giving it a uniform color, and this is a more cost-efficient solution than replacing the driveway.

Q. *During repair work, hydraulic oil from a forklift leaked onto an exposed aggregate driveway that was only 2 months old. The driveway wasn't sealed and is now stained by the oil. The stain occurred about a week ago. The driveway has a steep slope so any liquid used for cleaning will run down the slope. Also, we're in California and aren't permitted to use petroleum-based products or turpentine. Is there any way to remove the stain?*

A. Try using a poultice. Dissolve 1 pound, 6 ounces of sodium orthophosphate (also called trisodium phosphate or TSP) in a gallon of water. You can buy TSP at most hardware stores. Add enough finely ground calcium carbonate (also called whiting or agricultural lime) to make a thick paste. Agricultural lime is available at garden supply stores.

Spread the poultice over the stain and allow it to dry for a day if possible. Then brush off the dry paste with a stiff natural bristle brush and scrub the concrete with clear water.

A second possible solution is to sprinkle dry cement on the stain when the driveway is dry and the weather is sunny. Cover the cement with a glass plate. The sun's heat reportedly draws oil into the dry cement which can then be swept away.

As with any stain removal method, it's safest to try these two approaches on a small area first and observe the results.

Removing hydraulic oil stains

Q. *We heard that the greenish blue discoloration often found on concrete containing slag can be removed with a 3% hydrogen peroxide solution. Our experience suggests that this treatment may lead to scaling of early-age concrete. We found dilute solutions of vinegar to be effective. Some contractors have mentioned that di-ammonium citrate might also work. Any suggestions?*

A. *Removing Stains from Concrete*, published by The Aberdeen Group, also suggests using a 3% solution of hydrogen peroxide. A representative of a blast-furnace slag manufacturer has heard of using solutions containing as much as 30% hydrogen peroxide with no subsequent scaling. He has not heard of anyone using the other methods you suggest but believes they may be equally as effective. His experience suggests that exposure to sunlight and air is the best solution for discolored surfaces that are outdoors.

Cleaning concrete discolored from blast-furnace slag

Chapter 22

Testing

Q. *We are bidding a concrete project on which slump doesn't seem to be specified. In the specifications there is reference to ACI 301, "Specifications for Structural Concrete for Buildings," and to ASTM C 143, "Standard Test Method for Slump of Hydraulic Cement Concrete." The test method tells how to run the slump test but doesn't tell what slump is required for different types of construction. Is there any guideline for slump in this situation?*

A. Section 3.5 of ACI 301 deals with this situation. Unless otherwise permitted or specified, the concrete should be proportioned and produced to have a slump of 4 inches or less if it's to be consolidated by vibration. If it's consolidated any other way, the slump should be 5 inches or less. A tolerance of up to 1 inch above the maximum indicated is allowed for one batch in any five consecutive batches tested. Concrete of lower than usual slump may be used provided it is properly placed and consolidated. It's probably a good idea to confirm with the specifier that the lack of a slump specification isn't an oversight and that he'll permit the use of 4- or 5-inch-slump concrete.

Q. *We supply ready mixed concrete for many jobs that use the American Concrete Institute "Specifications for Structural Concrete for Buildings," ACI 301. We have many field-strength test results for all our standard mixes and we use these to show that proposed concrete proportions will produce the average strength required for a job. However, for some required strengths we don't have field test data and have to show results from suitable trial mixtures as discussed in paragraph 3.9.3.3 of ACI 301-89.*

Do the trial mixes have to be made and tested by an independent testing laboratory? We have a well-equipped quality-control laboratory and I'm a certified laboratory technician. We want to do the mixes ourselves, both to save money and to know for sure how the tests were conducted.

A. We find no requirement in ACI 301 that says trial mixes must be made and tested by an independent testing laboratory. That doesn't seem to be the intent of the specification.

In Chapter 16 on testing, however, paragraph 16.3.2 says that the testing agency designated in contract documents shall "review and check-test the contractor's proposed mixture design when required by the architect/engineer."

Q. *We've heard of the mat test for moisture in concrete. Where can we find out more?*

A. The moisture and surface-bonding characteristics of concrete slabs that are to receive floor coverings of rubber tile, solid vinyl tile, and vinyl sheet may be checked with the mat

Maximum permitted slump when slump isn't specified

Trial mixes by ready mix producers

Mat test for moisture

moisture and bonding test. The test may also be used before placing any resilient flooring on slabs from which paint, oil, adhesive, curing compound, or other coatings have been removed. The test is performed by placing 24-inch (600-mm) square linoleum or vinyl sheet mats on two adhesive bands—one a water-soluble adhesive, the other a water-resistant latex adhesive. The edges of the mats are taped to the floor. After 72 hours, the mats are removed and the adhesives examined. The presence of too much moisture will partially or completely dissolve the water-soluble adhesive, while the water-resistant adhesive will be stringy with little bond. If after further drying and retesting, moisture in the slab still affects the adhesives, more moisture-resistant alternate flooring materials should be used. An ASTM Standard is being developed.

Field testing for cold-weather concrete

Q. *We build residential basements and tried to work later in the season this year, but were concerned about the concrete coming up to strength in cold weather. Is there a field test that we can use to measure in-place strength?*

A. Several field tests for determining in-place strength are standardized by ASTM and are done with equipment available commercially. These tests are described briefly in the Portland Cement Association's *Design and Control of Concrete Mixtures,* 13th edition, and in more detail in "In-place Methods for Determination of Strength of Concrete," ACI 228.1R-89. The ASTM C 900 pullout test is used for this purpose. It involves casting an insert in the concrete, then pulling it to failure with a calibrated device. This test measures stresses on a cone-like surface around the insert and provides a result that can be correlated with compressive strength. In some cases, the cone doesn't pull out and little or no patching is required.

The pullout device is fairly expensive—$3,000 to $4,000. Also, there are costs for testing to calibrate pullout and compressive tests for typical winter concrete mixes in your area. We know of two companies that sell pullout testing devices in the United States:

Germann Instruments Inc.
8845 Forest View Rd.
Evanston, IL 60203
(847-329-9999)

James Instruments Inc.
3727 N. Kedzie
Chicago, IL 60611
(800-426-6500; in IL: 773-463-6565)

Chloride content by weight of concrete

Q. *For corrosion protection, ACI 318-95, "Building Code Requirements for Structural Concrete," specifies maximum water-soluble chloride ion in concrete as percent by weight of cement. What are acceptable levels when measuring chloride-ion content in existing concrete-specifically, by weight of the concrete sample?*

A. For an exact answer you need to know the cement content and unit weight of the concrete. For instance, if the cement content is 600 pounds per cubic yard, a maximum allowable chloride-ion content of 0.15% by weight of the cement corresponds to 0.9 pounds of chloride ion per cubic yard. If a cubic yard of concrete weighs 3,800 pounds (about 142 pounds per cubic foot), the level would be 0.9/3,800, or 0.024%, by weight of the concrete.

Q. *On an average job, how often should slump and air content tests be made? How many test cylinders should be made?*

A. Specifications usually dictate the number of tests made. The ASTM "Specification for Ready-Mixed Concrete" (ASTM C 94) says slump, air content, and temperature tests shall be made at the time of placement at the option of the inspector as often as is necessary for control checks. ASTM C 94 also says that these tests shall be made when specified and always when strength specimens are made.

Chapter 16 of "Specifications for Structural Concrete" (ACI 301-89) tells how often to make tests. Determine slump for each strength test and whenever consistency of concrete appears to vary. Determine air content of normal-weight concrete whenever a strength test is made.

ACI 301 also says to make at least one strength test for each 100 cubic yards, or fraction thereof, of each mixture design of concrete placed in any one day. When the total quantity of concrete with a given mixture design is less than 50 cubic yards, the strength tests may be waived by the engineer if, in his or her judgment, adequate evidence of satisfactory strength is provided.

Make three test specimens from each sample. Two are tested at 28 days for acceptance and one is tested at seven days for information.

How many concrete control tests are needed on an average job?

Q. *Is there a publication that explains standard aggregate tests in an easily understood manner? The ASTM standards for specific gravity, abrasion resistance, sieve analysis, and similar tests are hard reading.*

A. An American Concrete Institute publication, *Aggregates for Concrete* (El-78), should help you to better understand the ASTM test methods. Order this 60-page booklet from the American Concrete Institute, P.O. Box 9094, Farmington Hills, Mich. 48333.

More information about aggregate tests

In the June 1991 *Concrete Construction* Problem Clinic, a reader asked if there was a publication that explained standard aggregate tests in an easily understood manner. We recommended *Aggregates for Concrete*, published by the American Concrete Institute. Bryant Mather of the U.S. Army Corps of Engineers Waterways Experiment Station mentioned another helpful book that's published by ASTM. It's STP 169B, *Significance of Tests and Properties of Concrete and Concrete-Making Materials*. This book can be ordered from ASTM, 188 Barr Harbor Drive, West Conshohocken, PA 19428

Explaining aggregate tests

Q. *Before placing the concrete for the last pier of a drilled-pier foundation job, the foreman decided to add water to the ready mix truck. The inspector didn't like the looks of the watered-down concrete and took test cylinders that represented that one pier. The specifications call for a 28-day strength of 3000 psi. After the lab broke the seven-day cylinders, the cylinder from the pier with added water broke at 1980 psi. The other seven-day cylinders were as high as 2620 psi. The engineer is concerned that the concrete will not meet the specified strength. I realize that adding the water was the wrong thing to do, but I don't want to remove the pier if it is of adequate strength. Will it reach the specified 3000 psi?*

A. As this case shows, it is often useful to extrapolate 28-day strengths from seven-day strengths. Of course, the amount of strength gain varies between the seven-day and the 28-day tests. Cement type and curing conditions are two factors that affect the amount of strength gain to be expected.

Relationship between seven-day and 28-day strengths

Concrete, by Mindess and Young, gives a general rule: The ratio of 28-day to seven-day strength lies between 1.3 and 1.7 and generally is less than 1.5, or the seven-day strength is normally between 60% to 75% of the 28-day strength and usually above 65%. The cylinder that broke at 1980 psi is 66% of the specified 3000 psi. According to Mindess and Young's rule, it should meet the specified strength at 28 days.

Most likely, the mix wasn't designed for 3000 psi but for a higher compressive strength to account for variability. By adding the additional mix water you raised the water-cement ratio which, in turn, reduced the strength. The piers placed before the water was added will probably have strengths higher than the specified 3000 psi. The pier in question, however, will most likely meet the specified strength.

If after 28 days the cylinders still do not meet specified strength, take cores to verify the strength before implementing a costly pier removal.

Cap cylinders before curing them?

Q. *At our precasting plant we put sulfur caps on test cylinders the day after they're cast but before they're placed in the moist room. Is there anything wrong with doing this?*

A. ASTM's "Standard Practice for Capping Cylindrical Concrete Specimens" (ASTM C 617-87) doesn't directly address this practice. Section 5.2 says hardened specimens which have been moist cured may be capped with high-strength gypsum plaster or sulfur mortar. It doesn't say, however, that specimens which haven't been moist cured can't be capped with sulfur mortar.

If cylinders capped with high-strength gypsum plaster are stored in a moist room, the caps must be protected from dripping water and moist room storage must not exceed 4 hours. That's because water may weaken the caps. Cap weakening isn't a concern with sulfur caps because they aren't soluble in water.

You do have to be sure that cylinders don't dry too much after they're stripped from the mold and capped. Guard against this by wrapping the stripped cylinders in wet burlap before and after capping until they're placed in the moist room.

What's infrared thermography?

Q. *We're repairing a bridge deck that requires several partial-depth patches. The engineer recently handed me a computer printout of the bridge deck that showed the delaminated areas. I asked him how the delaminated areas were detected, and he said that one method they use is called infrared thermography. What is this technique and how does it work? Is it accurate?*

A. Infrared thermography is a nondestructive testing technique that has proven to be accurate and efficient in locating voids, delaminations, and other defects in concrete structures. It's based on the principle that heat flows more slowly through voids and delaminations (air) than through solid concrete. These changes in heat flow cause localized differences in surface temperature. By measuring the surface temperature under conditions of heat flow, the location of delaminations can be determined.

During daylight hours, the concrete surface above delaminations is warmer than the surface above sound concrete. That's because heat from the sun takes longer to move deeper into concrete because it moves slower through the delaminated area. The opposite is true at night—the concrete surface over delaminated areas is cooler. Because of this, the best times to perform the inspection are two to three hours after sunrise or sunset. Both are times of rapid heat transfer.

The procedure used to collect the data is outlined in ASTM D 4788, "Test Method for Detecting Delaminations in Bridge Decks Using Infrared Thermography." To measure surface temperatures and determine delaminated areas, engineers use high-resolution infrared thermographic scanners. The scanners' optical systems are transparent only to short- or medium-wave infrared radiation. They can detect temperature variations as small as 0.4° F.

The scanner head and an accompanying video camera are mounted on a vehicle at a height sufficient to allow a minimum image width of 14 feet. Scanning rates as great as 1.2 million square feet per day are possible. The resulting data can be displayed as pictures with areas of differing temperatures designated by differing gray tones in a black and white image, or by various hues on a color image. A wide variety of computer equipment is used to facilitate data recording and interpretation.

Reference

1. V. M. Malhotra and N. J. Carino, *Handbook on Nondestructive Testing of Concrete*, CRC Press Inc., 1991.

Q. *We produce ready-mixed concrete for many jobs requiring documentation that the proposed concrete proportions will produce average strengths specified for the job. The documentation includes determination of a standard deviation based on at least 30 consecutive field strength tests. Can we use 4x8-inch cylinder results for this determination? The smaller cylinders are easier to handle, and for high-strength concrete they don't require a large testing machine.*

A. We don't advise using the smaller cylinders for this purpose. ASTM's "Standard Practice for Making and Curing Concrete Test Specimens in the Field" (ASTM C 31-90) seems to prohibit smaller cylinders. In Section 5.1 the practice states that unless required by the project specifications, cylinders smaller than 6x12 inches shall not be made in the field.

There's another good reason for basing the standard deviation calculation on 6x12-inch cylinders. Smaller cylinders generally yield higher strengths but the test results also are more variable. This would increase the calculated standard deviation causing the required average strength (and cement costs) to be higher.

Can 4x8-inch cylinders be used for producer's standard deviation determination?

Q. *On my last project, the inspector rejected a concrete truckload if one slump test failed to meet the specifications. On previous jobs, I've had inspectors take two slump tests before rejecting the concrete. Which method is right, and is there a reference document relating to the answer?*

A. ASTM C 94, "Standard Specification for Ready-Mixed Concrete," states: "If the measured slump or air content falls outside the specified limits, a check test shall be made immediately on another portion of the same sample. In the event of a second failure, the concrete shall be considered to have failed the requirements of the specification." ASTM C 94 is referenced on most projects, since it is incorporated into the ACI 318 Building Code and into other national building codes.

ASTM C 143, "Standard Test Method for Slump of Hydraulic Cement Concrete," also requires making a second slump test on a second portion of the sample "if a decided falling away or shearing off of the concrete from one side or portion of the mass occurs." The standard further states that "if two consecutive tests on a sample of concrete show a falling away or shearing off of a portion of the concrete from the mass of the specimen, the concrete probably lacks necessary plasticity and cohesiveness for the slump test to be applicable."

Thus, using two slump or air tests before rejecting the concrete is the correct procedure.

Can concrete be rejected after one slump test?

Curing lightweight-concrete test cylinders

Q. *For the first time in years we have a job calling for lightweight concrete, but we haven't kept up-to-date on specifications for testing. The aggregate supplier told us to air dry the test cylinders after a seven-day moist cure. Is this correct?*

A. According to ASTM C 192, "Making and Curing Concrete Test Cylinders in the Laboratory," test cylinders for structural lightweight concrete can be moist cured from the time of molding until the time of testing, or the procedures of ASTM C 567, "Unit Weight of Structural Lightweight Concrete," can be followed. That specification calls for a seven-day cure at 60° to 80° F, without moisture loss from the specimens, followed by 21 days at 50% ± 5% relative humidity and 73.4° F ± 2°. These conditions are specified to determine the air-dry unit weight of lightweight concrete, and are said to closely approximate the "in-service" moisture conditions for a large part of the United States.

The air drying that you describe would be acceptable under the ASTM standards, but you may wish to verify that this is permitted in the contract documents for your project.

Can rebound hammer results determine safe form-stripping time?

Q. *We're on a job that requires a fast forming cycle. If we calibrate our rebound hammer for the concrete being used on the job, are the hammer test results accurate enough to determine when forms can be safely stripped?*

A. The one study we found says no. In the study, researchers computed within-test variability of the rebound hammer test at concrete test ages of one to three days. They also studied the ability of the test to determine early-age strength development of concrete for form removal purposes. The rebound tests were performed on plain concrete slabs. Companion cylinders and cores taken from the slabs were tested in compression.

After analyzing the test data and finding a large within-test variation, the researchers concluded that the rebound hammer wasn't a satisfactory method for reliably estimating early-age concrete strength development.

Reference

G.G. Carette and V. M. Malhotra, "In-Situ Tests: Variability and Strength Prediction of Concrete at Early Ages," *In-Situ/Nondestructive Testing of Concrete*, ACI SP 82, 1984, p. 111.

Chain drag delamination surveys

Q. *In the article "Shear Collars Save a Parking Garage" (Concrete Construction, October 1988), the authors refer to a chain drag delamination survey. This method seems to be much faster than the traditional method of tapping with a hammer to detect delaminations. Do you have specific details on the method?*

A. Gary Klein of Wiss, Janney, Elstner replies. The method simply involves dragging lengths of chain across the top of a concrete surface. A distinctly hollow, drum-like sound is heard when delaminations are encountered. Chain drag surveys are particularly effective in locating shallow delaminations on uncovered decks. The method can't be used to locate delaminations in the concrete below bituminous wearing surfaces, although debonding between wearing surface and underlying concrete can generally be detected by a chain drag survey.

A hammer or steel rod can be used to conduct a similar survey on vertical and overhead surfaces, although it's generally not practical to survey the entire underside of a bridge deck or parking deck.

Automated acoustical sounding devices also are available. We have found, though, that the chain drag survey is more cost efficient and reliable. In all cases, however, you should verify that a delamination is present by core drilling.

The locations of delaminations are typically marked on the deck surface and plotted on a plan view sketch of the bridge deck or parking structure. The percentage of delaminated area can then be determined. Quantities of repair materials used substantially exceed quantities predicted on the basis of delamination area measurements. That's because some incipient delaminations aren't detected by the chain drag survey. Also, some sound concrete has to be removed at the perimeters of the delaminated areas.

For more information about delamination surveys, see the article by Bill Perenchio entitled "The Condition Survey" in *Concrete International,* January 1989, page 59. For a copy of the article, contact the American Concrete Institute, P.O. Box 9094, Farmington Hills, MI 48333.

Q. *What's a reasonable frequency for testing fresh concrete?*

A. "Specifications for Structural Concrete," ACI 301-96, requires obtaining at least one composite sample for each 100 cubic yards, or fraction thereof, of each design mixture of concrete placed in any one day. Composite samples are obtained in accordance with "Standard Practice for Sampling Freshly Mixed Concrete," ASTM C 172. Strength tests are conducted on each sample, but when the total quantity of concrete with a given design mixture is less than 50 cubic yards, the strength tests may be waived by the architect/engineer if, in his or her judgment, adequate evidence of satisfactory strength is provided.

Slump tests are required for each composite sample and whenever consistency of the concrete appears to vary. If air entrainment is required, air-content tests are also required for each composite sample.

Where concrete will be exposed to deicing salts, air-content tests must be made on samples from the first three batches in the placement and until three consecutive batches have air contents within the specified range. After that, every fifth batch can be tested until a test isn't within the specified range, at which time testing of each batch must be resumed until three consecutive batches have air contents within the specified range. Additional tests may be performed as necessary for control. These air-content tests may be taken on composite samples or on samples from the batch taken at any time after discharge of 2 cubic feet of concrete.

Q. *We're on a job that requires the microsilica concrete to be less than 600 coulombs chloride permeability at 28 days for laboratory-cured specimens and at 80 days for field-cured core samples, as measured in accordance with AASHTO T 277-83. There's a pay reduction of almost $5 a square foot if this isn't met. We've seen specifications before for laboratory specimens, but never for core specimens. How will core specimens affect rapid chloride permeability results?*

A. It has been our experience to see two coulomb values one specified for lab-cured concrete and another for field-cured concrete. For a 600 coulomb lab value, some engineers use a 1,000 coulomb field value.

One problem with requiring the same value for both specimens is maturity of the field concrete. The specifications relate the curing of the field concrete with the laboratory concrete by waiting an additional 52 days. In cold weather, however, 52 days might not be the appropriate waiting time. The maturity method can be used to indicate when the field concrete equals the same maturity as the lab concrete. The maturity concept, however, assumes equal moisture

How often should fresh concrete be tested?

Chloride permeability testing of field-cured concrete

content. Depending on the field curing procedure, it's unlikely to provide the same moisture conditions as an ASTM-specified curing room.

Consolidation also is an important issue. Field consolidation rarely will be equal to or better than the consolidation of a cylinder made in accordance with ASTM. Research by the Portland Cement Association indicates that a reduction in consolidation from 100% to 90% increases the amount of chloride ions passed by 100%.

Another problem with requirements to check chloride permeability of field specimens at 80 days is a possible delay in acceptance and payment for a concrete pour. To help ensure timely payment, work with the project engineer to obtain an appropriate chloride permeability for a field-cured core at 28 days. This value should reflect the differences in field versus laboratory concrete.

Curing cylinders by immersion in water

Q. *Is it permissible to put concrete test cylinders in a tub of water to cure them?*

A. For initial curing, test cylinders can be immersed in saturated limewater at 60° to 80° F immediately after molding if they're not cast in cardboard molds or molds that expand when immersed in water. After 24 ± 8 hours they should be immersed in water at 73.4 ± 3° F until they're tested.

If the cylinders are in cardboard molds, prevent loss of moisture by covering them and keep the temperature at 60° to 80° F. Then they can be removed from the molds and immersed in saturated limewater at 73.4 ± 3° F.

Immersion in water provides the best possible curing environment. This is particularly important when the concrete has a low water-cement ratio. Immersion is also a nearly foolproof way of curing. If moist room fogging nozzles clog or there's a temporary loss of water pressure, cylinders in the moist room could dry, reducing strength. Immersion eliminates that concern.

Small-diameter cores more likely to yield low strengths

Q. *When there's a question about in-place strength of concrete as a result of low cylinder test results, cores anywhere from 1½ inches to 4 inches in diameter are taken. Is there any disadvantage in taking the smaller cores?*

A. We think you're better off drilling larger diameter cores. Here's why. Variability in core strength increases as the core diameter decreases. That means small-diameter cores have a greater chance of yielding a core test result that falls below the 0.75 f'_c lower limit given in paragraph 17.3.2.3 of "Specifications for Structural Concrete for Buildings" (ACI 301-89).

Using the break-off test to measure in-place strength

Q. *The Problem Clinic in the June 1993 issue of* Concrete Construction, *described the use of pull-out tests to determine the in-place strength of concrete. What about the break-off test?*

A. The break-off test also can be used to measure in-place strength of concrete in the field. In North America, there is less experience with this method, compared to pull-out tests, probe penetration, and the rebound hammer. The break-off test has been available commercially in the United States since 1986. The test procedure has been standardized as ASTM C 1150-90, "Standard Test Method for The Break-Off Number of Concrete." Reportedly, the device has been used widely in Europe for testing pavements and slip-formed structures.

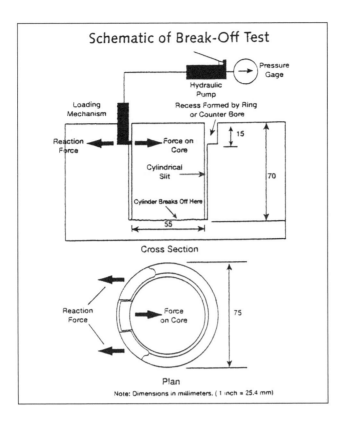

Schematic of Break-Off Test

The break-off test measures the force required, or break-off number, to fracture a small cylinder formed by reusable sleeves inserted in fresh concrete, or later cut by a special diamond drill bit and counterbore. Typically, break-off numbers from five individual tests are averaged. The manufacturer provides a generic correlation between the break-off number and results of compressive-strength tests, but this relationship is best developed for specific local combinations of concrete materials. Results of the break-off test are said to correlate better with concrete tensile strength than those obtained by pull-out tests.

In the United States, the break-off tester is available from: SDS Co., P.O. Box 844, Paso Robles, CA 93447 (805-238-3229; fax 805.238.3496).

Q. *We placed a concrete retaining wall that was 1 foot thick, 12 feet high, and over 300 feet long. The testing lab made one set of cylinders during the placement. Breaking of the 28-day cylinders showed low concrete strengths. The engineer would like to take cores from the wall to investigate the strength. Where in the wall should cores be taken?*

A. Coring is generally necessary when cylinder strengths are unacceptably low. Therefore, core strengths should be obtained from the in-place concrete that corresponds to the low cylinder strengths. Unfortunately, this often is not the case in practice.

ACI 318, "Building Code Requirements For Reinforced Concrete," requires the cores to be removed from the area in question. This means finding the suspect concrete from the truckload from which the cylinders were taken. If accurate placement records are available, the area of presumed low strength can be located and cores taken. If accurate placement records are not available, use a nondestructive testing technique to locate the area of suspected low strength.

These procedures satisfy ACI 318 requirements, but other locations may need testing. The

Where to remove cores when investigating low cylinder strengths

truckload of concrete from which the cylinders were taken may represent the concrete quality in other trucks as well. Using a nondestructive technique to locate the suspect truckload of concrete provides a comparison for locating other potential low-strength areas.

Methods for determining in-place flexural strength

Q. *We are placing a concrete pavement and are using beams to determine the adequacy of the in-place flexural strength. The required 28-day flexural strength is 650 psi. Unfortunately, during one day's placement, we forgot to notify the testing lab to come to the site to make the beams. Can we remove cores from the pavement and test them in compression to show that the concrete meets specifications? If so, what is the correlation between compression and flexural strengths?*

A. There are three possible methods of removing a concrete sample and testing it to estimate in-place flexural strength. The method you suggested is one way. Typically, the flexural strength of concrete is 11% to 18% of the compressive strength. Check the original mix design for the project. Usually the testing lab will make and test cylinders along with the beam specimens. This may give you the necessary correlation between compression and flexural strengths at 28 days.

Another method is to remove a core and test its split tensile strength. Split tensile strength can also be correlated with flexural strength. The Federal Aviation Administration in Engineering Brief No. 34, "Referee Testing of Hardened Portland Cement Concrete Pavement," provides a correlation between split tensile and flexural strength: Flexural strength = 1.02 x (split tensile strength + 200 psi). It's usually best, however, to correlate split tensile and flexural strengths with the materials used for each specific concrete mix rather than the generic formula provided by the FAA.

ASTM C 42, "Standard Test Method for Obtaining and Testing Drilled Cores and Sawed Beams of Concrete," provides a procedure for removing beams from in-place concrete by sawing, but this method isn't recommended. Test results of beams removed in this manner have been 30% lower than laboratory-cured beams. The sawing operation damages beams significantly and makes interpretation of the final results questionable.

Before you remove a core and test it in compression or split tension, check with the engineer to make sure he approves of the procedures and the strength correlation. A lot of money can be wasted on tests that are considered invalid. Make sure all parties agree on the following:
- Number of tests
- Location of test specimens
- Method of specimen removal
- Size and shape of test specimens
- Curing environment and duration
- Test methods
- Strength correlation
- Acceptance or rejection of final results

Cement fineness and how it is measured

Q. *What effect does the fineness of a cement have on the properties of both fresh and hardened concrete? Also, how is cement fineness measured?*

A. In general, the finer a cement, the greater will be the rate of hydration. This, in turn, leads to a higher rate of strength gain and a higher rate of heat generation. Because hydration takes place at the surface of cement particles, the finer a cement, the more completely it will hydrate. Cement particles larger than 75 microns may never hydrate completely. In addition, increasing fineness tends to decrease bleeding. However, a high fineness increases the amount of water needed for workability in non-air-entrained concrete. This can result in increased drying shrinkage. High cement fineness also reduces the durability of concrete to freeze-thaw cycles.

It is generally not recommended to use sieves to measure cement fineness because the fineness of most cement particles tend to clog very fine mesh sieves. The general practice is to measure fineness using specific surface area. Two methods of determining fineness are recognized by ASTM: the Wagner turbidimeter method (ASTM C115) and the Blaine air-permeability test (ASTM C204).

Q. *We recently supplied controlled low-strength material (CLSM) for backfill. Specifications called for a maximum 28-day compressive strength of 200 psi to allow excavation by backhoe at a later date. We designed for a strength between 50 and 150 psi, but cylinder breaks indicated compressive strengths exceeding 200 psi. Why would these breaks be so much higher than our design strengths? And is there a chance that the in-place strength is lower than the cylinder breaks indicate?*

A. We can think of a few materials-related reasons for the high test breaks. If you normally use a Class F fly ash in the CLSM but switched to a Class C ash, 28-day strength would increase. If your design mix had a high entrained-air content, but air was lost in transit or when test cylinders were made, the decreased air content would increase strength. Finally, if there was leftover concrete in the trucks when the CLSM material was batched, a higher-than-normal cement content and amount of coarse aggregate could account for some strength gain.

At least some of the higher-than-expected cylinder strength might also result from using a testing machine that isn't sensitive enough. Standard concrete testing machines shouldn't be used to test CLSM since the dials on many are readable only to the nearest 250 pounds.

Samples of CLSM are often fabricated and tested in accordance with ASTM D 4832, Strength Test Method for Preparation and Testing of Soil-Cement Slurry Test Cylinders. This test method states, "Since the compressive strength of soil-cement slurry cylinders will typically be 100 to 300 psi, the testing machine must have a loading range such that valid values of compressive strength can be obtained in accordance with ASTM C 39." ASTM C39, "Standard Test Method for Compressive Strength of Cylindrical Concrete Specimens," states, "In no case shall the loading range of a dial be considered to include loads below the value that is 100 times the smallest change of load that can be read on the scale." If you tested 6x12-inch cylinders, it would take only a 2,830-pound load to generate a 100-psi stress. That means the dial readability must be 2,830/100 or about 25 pounds. Thus if you're reading to the nearest 250 pounds, it would be easy to have a dial reading error of about 10% for a 100-psi material.

It's likely that the in-place strength will be higher than the cylinder strength. That's because the cylinders are made in impervious molds, and little water is lost up to the time of testing. At the jobsite, some water migrates out of the CLSM and into the surrounding soil. Because of this, the water-cementitious materials ratio is likely to be lower—and the strength higher—for in-place CLSM. There's also some long-term strength gain for the in-place material. If excavation is done a year after placement, there's a good chance that the CLSM in-place strength will be higher than the 28-day cylinder strength.

Q. *We're supplying concrete for a tilt-up project that requires concrete with a 3500-psi (25 MPa) 28-day compressive strength. In addition to making standard test cylinders, the tilt-up contractor is also making beams that are stored in the field before they're tested. Because of nonstandard curing conditions for the beams, should we be concerned about being held accountable for the beam strengths?*

Tilt-up contractors sometimes cast test beams to determine modulus of rupture.

A. No. The beams and cylinders are made for two different purposes. The tilt-up contractor's erection manual gives a strength that in-place concrete must attain before the panels can be lifted. This is not the same as the 28-day strength specified for standard-cured cylinders. Typically the required in-place compressive strength is around 2500 psi (18 MPa), which is usually reached in five to seven days.

Some contractors, however, use beam tests to determine when the panels are ready for lifting. When cylinders are made, the contractor's testing laboratory also casts companion beams that are left at the jobsite so they're exposed to the same conditions as the tilt-up panels. Before lifting, laboratory personnel test the beams and calculate the modulus of rupture. Generally, panels are designed to be lifted when the modulus of rupture reaches 500 psi.

You're responsible only for the strength of the cylinders cured as described in ASTM C 31, Section 9.2, "Standard Curing." The beam strength is affected by jobsite conditions such as temperature and curing methods over which you have no control.

Flexural strength testing

Q. *We supplied concrete for a recent pavement job where the acceptance test was ASTM C 78, flexural beams with third-point loading. For some time, the contractor was experiencing low results. We became involved because the mix provided was in question. The inspector mentioned allowing ASTM C 293, flexural testing with center-point loading, would produce higher-strength results. Is this true? If it is, can we substitute center-point for third-point loading?*

A. There are two different ASTM standard test methods used to calculate the modulus of rupture, the flexural strength of concrete. The main difference between these tests is the location where the load is applied (See figure). In center-point loading (ASTM C 293), the load is applied at the midspan of the beam. During third-point loading (ASTM C 78), the load is applied at the third-points along the length of the test span. The third-point loading arrangement carries maximum stress along the middle third of the test specimen, while the center-point loading carries maximum stress only at the location where the load is applied.

For a given beam size, center-point loading strengths are normally higher than those from third-point loading. The weakest plane is more likely to be encountered in the third-point configuration versus the center-point method, thus strength test results 15% higher by the center-point loading configuration are not unusual. However, the center-point loading configuration also experiences higher variability than the third-point test. This may affect overdesign when trying to meet a specified strength.

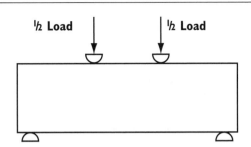

ASTM C 78 – Third-Point Loading, half the load is applied at each third of the span length. The modulus of rupture so measured is lower than by C 293. The maximum stress is present over the center third.

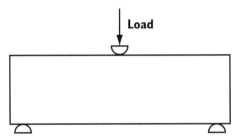

ASTM C 293 – Center-Point Loading, the entire load is applied at center span. The modulus of rupture will be greater than C 78. The maximum stress is present only in the center of the span.

Unfortunately, the contractor or ready mix supplier normally doesn't specify which test will be used for acceptance. The project specifications typically state the test method required. Which flexural test used seems to depend largely on geographical location, with some states relying on one method over the other. ASTM C 293 states that it is not an alternative to ASTM C 78.

Flexural tests are extremely sensitive to specimen preparation, handling, and curing procedures. For this reason, many state highway agencies are turning to the use of cylinders in lieu of beams for acceptance testing (References 1 and 2). Contractors and inspection agencies are more familiar with the traditional cylinder compression test for control and acceptance of concrete. With proper testing, a relationship between flexural and compressive strength can be obtained. Suggest this as an alternative on the next project which specifies flexural strength.

References

1. Meininger, R.C. and N.R. Nelson, "Concrete Mixture Evaluation and Acceptance for Air Field Pavements," National Ready Mixed Concrete Association Publication No. 178.

2. NRMCA, "Flexural Strength of Concrete," Concrete-in-Practice Sheet No. 16, National Ready Mixed Concrete Association, 1984.

3. P. Carrasquillo, "Concrete Strength Testing," Significance of Tests and Properties of Concrete and Concrete Making—Making Materials, ASTM STP 169C, 1994, pp. 123-139.

Estimating concrete slump

Q. *Are there any guidelines in "guesstimating" the slump of fresh concrete our ready mix truck drivers can use?*

A. The National Ready Mixed Concrete Association Truck Mixer Driver's Manual gives the following ways to estimate concrete slump:

1. Observing the concrete through the charge hopper.
2. The sound of the concrete rolling in the drum.
3. The load on the engine necessary to turn the drum.
4. Some mixers have hydraulic pressure meters. The "slump meter" measures the resistance of the drum as it turns the concrete. Slump meters should be calibrated against a properly run slump test once a week. Very stiff concrete will have a high resistance to turning and wet concrete will roll with less resistance.

Do not place too much emphasis on any individual estimating technique. Different mix designs, aggregates, size of load, and type of truck mixer will affect how the concrete mixes in the drum.

The following chart can be used to estimate the slump when discharging a load:

Observed Condition	Approximate slump	Discharge characteristics
Very stiff	1 inch or less	Will not flow, crumbly consistency
Stiff	2 to 3 inches	Needs help to flow down chute
Medium	4 to 5 inches	Flows well, piles on subgrade
Wet	7 inches or more	Very fluid, piles slightly or not at all

Testing flowable fill

Q. *Our ready mix company is thinking about marketing flowable fill in our area. However, one question has come up: How is flowable fill commonly tested?*

A. Flowable fill, or controlled low strength material (CLSM), is finding increasing use as backfill, structural fill, and for filling wells, abandoned tanks, and other underground structures. The National Ready Mixed Concrete Association (NRMCA) provides the following guidelines for testing flowable CLSM mixtures:

- Sample and remix the sample according to ASTM C 172.
- For compressive strength testing, use 6x12-inch plastic cylinder molds; fill to overflowing, then tap the sides lightly. Cover and cure the cylinders in the molds until the time of testing (or at least 14 days). Strip the mold carefully using a plastic knife to cut the mold off. Be aware that capping the cylinders with sulfur compounds can break CLSM materials. Neoprene caps have been used, but high-strength gypsum plasters seem to work best.
- Slump testing is not recommended because a very wet consistency is required for proper self-leveling. ASTM C 939 for flow of grout can be used by wet screening to remove coarse particles. An efflux time of 10 to 26 seconds through a special flow cone with a ½-inch discharge tube has been used.
- Unit weight and yield according to ASTM C 138 are done by normal procedures.
- Air content (if air entrainment is used) is measured by pressure meter according to ASTM C 231.
- Penetration resistance tests such as ASTM C 403, may be useful in judging the setting and strength development up to a penetration resistance number of 4,000 (roughly 100 psi compressive cylinder strength).
- Density tests are not required because flowable fill becomes rigid after hardening.

Reader response: Your answer to the question about testing flowable fill (July 1995, p. 609) may be misleading to some. Even mentioning neoprene caps, in my opinion, may lead some astray.

ASTM C 1231, "Standard Practice for Use of Unbonded Caps in Determination of Compression Strength of Hardened Concrete Cylinders," states: "Unbonded caps are not to be used for acceptance of testing concrete with compressive strength below 1500 psi or above 7000 psi." Unless another standard is applied when testing slurry mixes, ASTM C 1231 precludes pad usage in flowable fill strength ranges. Arguably, flowable fill is not concrete, but C 1231 seems to be the only applicable standard.

When samples of flowable fill are fabricated and tested by ASTM D 4832, "Strength Test Method for Preparation and Testing of Soil-Cement Slurry Test Cylinders," and capped with high-strength plaster gypsum, the results are usually very dependable once everyone has a little experience and confidence using these methods.

Dick Miller, Manager Quality Control, CalMat Co., San Diego

Q. *What is meant by yield of concrete and how is it determined?*

Measuring yield

A. Yield is the volume, measured in cubic feet or cubic yards, of concrete produced from a mixture of known amounts of cement, water, aggregates, and admixtures.

Concrete is batched by weight, but is sold by volume. Because of this, a unit weight test usually is conducted when it is necessary to calculate yield. By finding out how much a cubic foot of concrete weighs and dividing that number into the batch weights, the batch volume or yield is determined.

For example, if a 6-cubic-yard batch of concrete contains:

      3,156 pounds of cement
      7,056 pounds of sand (with 5% moisture)
     12,060 pounds of coarse aggregate
      1,230 pounds of mix water
     23,502 pounds total batch weight

and 1 cubic foot of concrete from the batch weighs 145.9 pounds, the batch's yield is:

     23,502/145.9 = 161.08 cubic feet

To determine the yield in cubic yards, simply divide 161.08 by 27 (the number of cubic feet in a cubic yard) to obtain 5.97 cubic yards.

Q. *A two-story motel was damaged by fire. Hollow-core panels on the second floor have exhibited color change and map cracking. Where can we find out what kind of tests are needed to determine structural soundness?*

Investigating fire-damaged hollow-core panels

A. We recommend hiring a consulting engineer with experience in evaluating fire-damaged structures. A listing of firms offering this specialized service starts on page 278 of the March 1991 issue of *Concrete Construction* magazine. The following three publications will increase your knowledge of the investigative methods used:

"Assessment of Fire-Damaged Concrete Structures and Repair by Gunite," Concrete Society Technical Report No. 15, 1978, The Concrete Society, 112 Windsor Rd., Slough, SL1 23A, England.

"Evaluation and Repair of Fire Damage to Concrete," SP-92, 1986, American Concrete Institute, P.O. Box 9894, Farmington Hills, MI 48333.

"Fire Safety of Concrete Structures," SP-80, 1983, American Concrete Institute, P.O. Box 19150, Detroit, MI 48219.

A paper by Armand Gustaferro in the last reference given discusses fire effects on precast, prestressed concrete construction as well as cast-in-place and post-tensioned concrete.

Coulomb testing raises questions

Q. *My state Department of Transportation is concerned about chloride ion penetration in concrete. It might specify a range of "coulombs" for approval of concrete and has bought some testing equipment. How can it test for chloride penetration? What is a coulomb?*

A. ASTM C 1202-91 "Test Method for Electrical Indication of Concrete's Ability to Resist Chloride Ion Penetration" gauges a concrete specimen's ability to resist penetration of chloride ions. Corrosion occurs more rapidly when significant amounts of chloride ions penetrate to the level of reinforcing steel. Reportedly, this test can reveal in two days what conventional tests (such as the AASHTO Standard T 259 "Method of Test for Resistance of Concrete to Chloride Ion Penetration") have required three months to measure. The trade-off versus more conventional test procedures is that the test is only an indirect measure of chloride penetration into the concrete.

The method is applicable to concrete for bridge decks, substructures and superstructures, marine and parking structures, and other corrosive environments. It provides data that can be compared to that from ponding-type tests.

C 1202 consists of passing an electrical current through the ends of a concrete core or cylinder for six hours. One end is immersed in sodium chloride, the other in sodium hydroxide. The total charge (or the number of coulombs) transmitted is a measure of the concrete's permeability to chloride ions. The "coulomb," named after French physicist Charles-Augustin de Coulomb, is the amount of electricity passed by a current of one ampere in one second: 1 ampere equals 1 coulomb per second. The letter Q is used to symbolize this measurement. A low value of coulombs passed during the test (for example, less than 1,000 coulombs over the 6-hour test period) is an indication that the concrete will resist penetration of chloride ions more than a concrete with a high number of coulombs passed during the test.

C 1202 recommends that numerical test results be used with caution, especially in applications such as quality control and acceptance testing. The standard recommends that qualitative descriptions of chloride ion penetrability be used in most cases. The results of two properly conducted tests by the same operator on concrete samples from the same batch could differ by up to 35%.

Chloride ion penetrability based on charge passed*

Charge passed, coulombs	Chloride ion penetrability
>4,000	High
2,000-4,000	Moderate
1,000-2,000	Low
100-1,000	Very Low
<100	Negligible

*ASTM C 1202-91

Q. *Our ready-mixed concrete company received a request for concrete with flexural strength of 670 psi. We have never had this kind of order before. Can you tell us what equivalent compressive strength such a concrete represents?*

A. This sounds like an order for paving concrete, which often is specified in terms of flexural strength, or modulus of rupture. The relationship between compressive and flexural strength varies for each concrete mix and depends on a number of factors, including size and type of aggregate.

A compression test squeezes the concrete, ideally failing the concrete through the aggregate. A flexural test pulls the concrete apart, with failure typically occurring at the bond between the cement paste and the aggregate. Thus, there is no direct relationship between concrete compressive and flexural strength.

According to the Portland Cement Association, the flexural strength of normal-weight concrete often is approximated as 7.5 to 10 times the square root of the compressive strength. ACI Committee 330 recommends using the following formula as an approximate relationship between compressive and flexural strength:

$$M_r = 2.3f_c^{2/3}$$

According to the formula, a flexural strength of 670 would be approximately equivalent to 5000 psi compressive strength. In addition to the formula, the ACI 330 chart shown here can be used as a quick reference. To optimize the use of your materials, the best method is to determine your own relationship between flexural and compressive strength specifically for the mix you will be using.

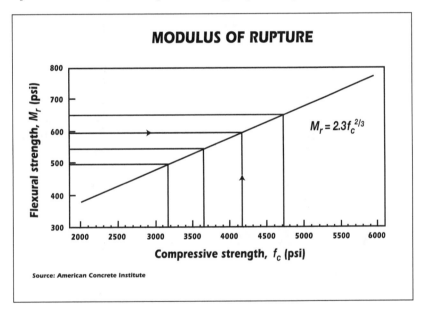

MODULUS OF RUPTURE

$M_r = 2.3f_c^{2/3}$

Flexural strength, M_r (psi) vs. Compressive strength, f_c (psi)

Source: American Concrete Institute

Q. *On a bridge repair, the engineer is requiring removal of all concrete containing more than 2 pounds of chloride ions per cubic yard. I've received chloride ion content reports from two materials testing labs. One reported the chloride ion content in percentage by weight of concrete, the other in parts per million. How do I convert these to pounds per cubic yard?*

A. One part per million equals 0.0001 percent by weight, so both labs are giving you information in a similar format. But to convert either result to pounds of chloride ion per cubic yard of concrete you need to know the density of the concrete. In the laboratory, this requires weighing a sample of the concrete and determining its volume as described in Paragraph 8 of ASTM C 1084.

If the chloride-content determination was based on powder samples from drilling, the lab wouldn't be able to provide density data. In this case, an assumed 3,800 pounds per cubic yard (about 142 pounds per cubic foot) for normal weight concrete can help you get an approximate conversion. As an example, let's say Lab A reports 0.0624 percent by weight of concrete and Lab B reports 580 parts per million (0.058 percent by weight).

Divide Lab A's reported value by 100 and multiply by 3,800.

Thus 0.000624 x 3,800 = 2.4 pounds per cubic yard. Using Lab B's 580 parts per million, 0.00058 x 3,800 = 2.2 pounds per cubic yard.

Testing cement uniformity

Q. *We have been experiencing variations in air content, slump, and concrete strength. Everything seems to stay consistent for a few weeks, and then there is a sudden change. We have been through our operation and haven't been able to locate any problems. We think that the cement might be responsible. What could be happening to the cement to cause this? Also, what low-cost options are available for us to determine if the cement is the source of the problems?*

A. Cement is cement is cement. How many times have you heard someone say that? Nothing is farther from the truth. One concrete producer asserts that cement is the single most variable material used in concrete. Although there are many possible reasons for your problems, cement could easily cause the variations in concrete you have experienced. Changes in the physical and or chemical properties of the cement can have a significant impact on concrete's air content, water demand, rate of slump loss, setting time, early-age strength, and later-age strength. Let's take a look at some of the reasons why cement properties vary over time and also some methods that a concrete producer can use to look at cement uniformity.

Variations in cement

All cement has some variations in physical and chemical properties that are a result of the manufacturing process. The amount of variation depends on the consistency of the rawmaterials used, the production quality, and the condition of the production equipment. Cement plants that have consistent raw materials, good production control, and well-maintained equipment generally can produce cement that has good uniformity. As a cement becomes less uniform, the properties of concrete produced with it become less consistent.

Color comparison test: Spoonfuls of cement from this week, last week, and last month are placed next to each other and pressed with a plastic or glass plate. The sample from the current week and the previous week are the same color, indicating that they are similar in chemistry. However, last month's sample is considerably darker in color, indicating that a substantial change in the cement has occurred.

Substantial changes in concrete properties can occur when there is a major modification of the cement's production process or chemistry. New production equipment may be put on line or the fuel type may be changed. Customers may demand quicker setting time or higher strength. The cement chemistry also may be modified to meet ASTM requirements for low heat of hydration, resistance to attack from high sulfate ground water, or use with alkali reactive aggregate. If a shortage occurs, cement supply typically will be supplemented by deliveries from a different source. In all of these situations, the chemistry will be different and the properties of the concrete can change.

Concrete producers should strongly urge the cement company to inform them when a change in chemistry or source will occur. Preparations can then be made to test and adjust the mixes instead of wondering why the concrete is behaving differently than it had the previous week.

Testing Cement

Although some concrete producers conduct standard cement tests as part of their quality assurance and quality control programs, purchasing the necessary equipment is an investment that most producers cannot justify. However, there are guidelines that all concrete producers should follow:

1. Cement should be sampled from bulk delivery weekly. Upon delivery, dig down into the cement in the truck about a foot or use a sample tube. Place about 8 pounds of cement in a large zip-lock bag. Make sure that the bag is sealed airtight and save it for at least two months. If you experience concrete problems that appear to be cement related, obtain larger samples more frequently. If the need for testing of the cement arises, do not part with the entire sample. A fairly complete set of field tests for physical and chemical properties should not require more than 2 or 3 pounds.

2. Compare color by taking spoonfuls of the current cement sample, last week's sample, and last month's sample and placing them on a counter so that each sample is in contact with the other two. Press a flat plastic or glass plate down on all three piles at once. If there is a color difference between the samples, something significant in the cement has changed. This test is particularly useful for determining when a major change in chemistry has occurred or when a different source of cement has been delivered.

3. Concrete producers should request ASTM C 917 reports from cement companies. The 917 report is a statistical evaluation of the uniformity of cement strength. When graphed, the data will show general trends of higher or lower strength over a period of time. Dramatic peaks and valleys show inconsistent material while a relatively flat curve indicates good uniformity. Plot strength results from a frequently used concrete mix on the same graph as the cement strength results. If a drop in concrete strength coincides with a valley in the 917 report, the cement is most likely responsible.

4. Another useful report available from cement companies is the cement mill certification. These certifications supply information about the chemical and physical properties of the cement. A monthly report usually is sufficient, but weekly or bi-weekly reports are preferred. The data supplied on these reports can be inserted into a spreadsheet and graphed similar to the data in the 917 report. This will yield useful information on the uniformity of the cement. Like the 917 report, changes in concrete performance can frequently be associated with changes in the data from mill certifications. Because concrete properties such as early or later strength, slump loss, air loss, setting time, and admixture compatibility are related to cement chemistry, the mill certification also can give a reasonable indication of how the cement will work in concrete.

Q. *Is there any way to determine, after concrete has hardened, whether a superplasticizer was added to the concrete? Is there a way to tell how much?*

A. Yes, it is possible to determine if a superplasticizer was used in concrete, but few laboratories are equipped to do the required tests. The process involves making a solvent extraction

Detecting superplasticizer in hardened concrete

and then doing an infrared analysis to determine if a melamine or phenol formaldehyde system is present.

In some cases, if you already know the specific identities of the chemicals used, it is possible to determine the amount of superplasticizer used.

Does position in structure affect rebound hammer results?

Q. *When using a rebound hammer to estimate the strength of in-place concrete, does it make a difference where you take the reading? For instance, will you get a lower reading near the bottom of a wall because of the dead weight of concrete above?*

A. Readings may vary depending on position in the structure, but the variations aren't caused by differences in dead weight.

Concrete near the bottom of an 8-foot-high wall feels a stress of only 8 psi caused by the dead weight of the concrete above (150 pounds per cubic foot x 8 feet = 1,200 pounds per square foot. 1,200 pounds per square foot/144 square inches per foot = 8 psi).

It's more likely that concrete near the top of the wall will have a lower rebound reading. That's because some of the bleedwater may not get to the surface before the concrete sets. This trapped water slightly lowers the strength.

Remember that the rebound hammer gives only a crude estimate of concrete strength, with an accuracy of perhaps 15% to 25%.

Chapter 23

Tolerances

Q. *We're bidding on a concrete floor job that requires a minimum F_F of 20 and a minimum F_L of 17. What do these numbers mean and how hard is it going to be to build to these specs?*

A. The F-numbers refer to flatness and levelness. F_F is an indicator of floor flatness or smoothness. F_L indicates floor levelness. The higher these numbers, the flatter or more level the floor.

An F_F 20 isn't a particularly flat floor, but you'll need to use an 8- or 10-foot straightedge to cut down high spots and fill in low spots during the finishing operations.

For a review of F-numbers and factors influencing floor flatness and levelness, see the article by Bill Phelan on page 5 of the January 1989 issue of *Concrete Construction*.

Understanding F-numbers for floor flatness and levelness

Q. *What does a specification for a floor flatness of ⅛ inch in 10 feet equal in F-numbers?*

A. Although there are no direct equivalents between F-numbers and straightedge tolerances, ACI 302, "Construction of Concrete Floors and Slabs," gives the following table of approximate values:

Floor-flatness number conversion

F-number	Gap under an unleveled 10-foot straightedge (Fraction of an inch)
F_F12	½
F_F20	⁵⁄₁₆
F_F25	¼
F_F32	³⁄₁₆
F_F50	⅛

Q. *Is there a published thickness tolerance for residential driveways? I'm a concrete consultant, and one of my clients is a homeowner who had a concrete driveway built. The contract specified a 5-inch thickness. After the concrete had cured, the owner cored it and found that the thickness was only 3½ inches. When the contractor was told this, his only response was that the driveway would perform well even though it wasn't 5 inches thick.*

A. Surprisingly, we couldn't find a thickness tolerance for residential driveways or even for pavements in "Standard Specifications for Tolerances for Concrete Construction and Materials"

Thickness tolerance for residential driveways

(ACI 117-90). The thickness tolerance for concrete slabs in section 4.4.1 of that document is +⅜ and -¼ inch, but it applies only to cast-in-place concrete buildings. Section 12, "Pavements and Sidewalks," gives only tolerances for lateral alignment of dowels and for level alignment.

We also found no reference to thickness tolerances in "Guide for Design and Construction of Concrete Parking Lots" (ACI 330R-92). But section 5.4 cautions that proper elevation control is critical because insufficient thickness due to poor grade control can be a serious deficiency.

State and local highway agencies generally publish thickness tolerances in their standard specifications. It's highly unlikely that any of these specs would permit a 1½-inch tolerance in a 5-inch-thick pavement.

There are thickness tolerances for driveways

Q. *Questions such as the one in the March 1997* Concrete Construction *Problem Clinic (p. 313), where a reader asks if there is a published thickness tolerance for residential driveways, bring out the importance of placing important construction parameters in easily accessible formats. There are recommended tolerances for thickness in "Guide for Design and Construction of Concrete Parking Lots" (ACI 330R-92). But evidently one has to know where to look, and that doesn't serve the occasional user very well.*

A. The last paragraph in section 4.2, "Subgrade Preparation," says: "Suggested tolerances for fine grading are no more than ¼ inch above or ½ inch below the design grade. Deviations greater than these tolerances can jeopardize pavement performance because small variations in thickness of thin pavements significantly affect load-carrying capacity." As the initial chairman of ACI Committee 330, I'm embarrassed that we did such a good job of hiding that important recommendation.

ACI Committee 330 is currently revising ACI 330R-92, and I shall recommend that a section on recommended tolerances be added to collect and tabulate the recommendations scattered throughout the report.

Richard O. Albright, Arsee Engineers Inc., Noblesville, Ind.

Tolerance on top-of-wall elevation

Q. *After pouring a cast-in-place wall, we discovered that the top-of-wall elevation doesn't match the elevation shown on the plans. What is an acceptable tolerance for top-of-wall elevation?*

A. "Standard Specifications for Tolerances for Concrete Construction and Materials" (ACI 117-90), doesn't call out this tolerance very clearly. Section 4, "Cast-in-Place Concrete for Buildings," doesn't mention walls. However, section 4.3.3, which covers level alignment, gives a ½-inch tolerance for lintels, sills, parapets, horizontal grooves, and other lines exposed to view. Tops of walls could conceivably be categorized as "other lines exposed to view," and would thus be subject to the ½-inch tolerance. However, any official interpretation of ACI standards must be made by the committee responsible for writing the standard.

Plumbness tolerance for walls

Q. *In the older edition of "Specifications for Structural Concrete for Buildings" (ACI 301-89), table 4.3.1 shows an allowable variation from plumb in the line of the wall as ¼ inch in any 10 feet of length and a maximum of 1 inch for the entire length. I've always thought this meant that a 10-foot-high wall could be no more than ¼ inch out of plumb to be within tolerance.*

No tolerances for formed surfaces are given in ACI 301-96, but the newer document refers to "Standard Specifications for Tolerances for Concrete Construction and Materials" (ACI 117-90). In ACI 117-90, section 4.1.1 on vertical alignment, the allowable variation for lines, surfaces, and arrises for heights 100 feet or less is given as 1 inch. No mention is made of ¼ inch in any 10 feet. Does this mean that a 10-foot-high wall can be 1 inch out of plumb and still be within tolerance if ACI 117-90 is the cited specification?

A. Any official interpretation of American Concrete Institute (ACI) standards must be made by the committee responsible for writing the standard. It appears that section 4.1.1 means a wall of any height up to 100 feet can be 1 inch out of plumb and still be within tolerance. In a subsequent reference to relative alignment in ACI 117-90, section 4.5.3 states:

"Formed surfaces may slope with respect to the specified plane at a rate not to exceed the following amounts in 10 feet:
  • Vertical alignment of outside corner of exposed concrete columns and control joint grooves in concrete exposed to view ¼ inch
  • All other conditions ⅜ inch."

This seems to introduce an allowable deviation from the specified plane of ⅜ inch in 10 feet, with no upper limit on the total amount. However, it's not clear to us how this relates to the requirements in section 4.1.1.

Elevated-slab thickness tolerance

Q. *We are working on a project where we placed a waffle-pan slab. Recently, we cored through the slab to install piping and conduit. The slab thickness above the pans was specified to be 4 inches. After analyzing the cores, we discovered that the slab is as thin as 3 inches in some sections and as thick as 4½ inches in others. What is the tolerance for this type of slab, and is this slab acceptable? If not, what should I do in lieu of ripping out the slab?*

A. American Concrete Institute Committee 117 report, "Standard Specifications for Tolerances for Concrete Construction and Materials," offers tolerances for cast-in-place concrete construction. For cross-sectional dimensions of members such as columns, beams, piers, walls (thickness only), and slabs (thickness only) the tolerances are as follows:

12 inches or less	+⅜ inch
	-¼ inch
More than 12 inches but not over 3 feet	+½ inch
	-⅜ inch
Over 3 feet	+1 inch
	-¾ inch

For a 4-inch-thick slab, this would limit the thickness to between 3¾ and 4⅜ inches, so this particular slab does not meet the required thickness after applying the tolerance.

Before planning on removing the slab, present this problem to the structural engineer. Possibly the 4-inch thickness was chosen because the designer wanted to keep nominal dimensions for ease of construction. The 3-inch-thick slab areas may be adequate to handle the necessary loads. The dimensions of the joists and the placement of reinforcement within them are often the critical elements in a waffle slab. The engineer may want to see if a reduction in load capacity is acceptable to the owner and meets building code requirements. Using it as built, it would obviously be more economical than ripping out the slab. In any case, the engineer should check the capacity of the floor in areas subject to large point loadings, such as those from heavy equipment or bookshelves.

Recommended floor slope and tolerance

Q. *We're designing the concrete floor for an auto dealership's service department. Trench drains are to be installed in the floor, but we don't know what slope is required to make sure that liquids flow into the drains. Also, what is a reasonable tolerance on the slope we decide upon?*

A. We could find no reference to recommended slopes when trench drains are used in a floor. The recommended slope for flat roofs is usually 1% (⅛ inch per foot) or 2% (¼ inch per foot) according to *The Professional Handbook of Building Construction*, published by John Wiley & Sons, New York. A 2% slope will probably ensure better drainage.

We could also find no slope tolerance for use when a floor slope is specified. ACI 117-90, "Standard Specifications for Tolerances for Concrete Construction and Materials," gives a level alignment tolerance of ¾ inch for top-of-slab elevation of slabs on grade. If the slab elevation at the trench drain is within ¾ inch of the plan elevation, and the slab top elevation farthest from the drain is also within ¾ inch of plan elevation, the slab would meet ACI 117 tolerances. This tolerance is useless, however, as a means for ensuring floor drainage.

Footing elevations

Q. *What is the tolerance for elevations on cast-in-place concrete footings?*

A. According to American Concrete Institute "Standard Specifications for Tolerances for Concrete Construction and Materials," ACI 117, Section 3.3, Level Alignment, the tolerance is ±½ inch for the top of footings supporting masonry, and +½ inch, -2 inches for the top of other footings

Tolerances for slab on grade thickness

Q. *What are the thickness tolerances applicable to slabs on grade?*

A. ACI 117, "Standard Tolerances for Concrete Construction and Materials," doesn't provide specific guidance on a thickness for slabs on grade. Generally, ACI 117 requires tolerances of +⅜ and -¼ inch for slabs up to 12 inches thick. But in an April 1989 *Concrete Construction* article, consultant Armand H. Gustaferro concluded that these tolerances were based on formed construction and are unrealistic for slabs on grade. Gustaferro described the results of measurements of 2,111 cores, representing more than 900,000 square feet of 4- and 6-inch-thick slabs. The slabs averaged 0.35 inch thinner than specified, and had a standard deviation of about 0.65 inch. He suggested that a reasonable specification would require that the average measured slab thickness exceed that shown on the drawings, and not more than 16% of the measurements be less than ⅜ inch thinner than the specified thickness. For a project typical of his data, the required average slab thickness would need to be 5¼ to 5⅜ inches to meet these criteria.

Gym floor flatness

Q. *We are bidding a job that has a slab-on-grade gymnasium floor with specified floor flatness of F_F 100 / F_L 50. Is this reasonable?*

A. It depends on the overall quality of the facility and the nature of the floor finish that will go over the concrete. This specification seems to be rather restrictive for ordinary facilities. We asked several floor contractors and consultants for their experiences and recommendations. They suggested F in the range of 35 to 50, and one thought that F_F 35$_F$/F_L 25 would be sufficient for most cases. The specifier on this project obviously is serious about obtaining a flat floor, but may not understand the costs. Work with the specifier and a prospective flooring installer to review performance needs. Consider the whole floor system, not just the flatness issue.

Q. *I serve as the technical troubleshooter for my company and answer many questions on why concrete cracks. One question that is asked regularly and is difficult to answer concerns what is an allowable crack width. Is there any literature or standard reference that discusses when a crack is too wide?*

A. Your question comes up on almost every concrete project. Unfortunately, there isn't just one easy answer because an allowable crack width is usually project specific. Also, don't forget that the answer can be approached from the perspective of the engineer, architect, owner, or contractor. An acceptable crack width on one project might be unacceptable on another project, such as one requiring an architectural finish.

ACI 224, however, does provide a general guide for crack widths at the tensile face of reinforced concrete structures for typical exposure conditions (see table). The committee has labeled the table "Tolerable Crack Widths," but notes that the table is intended to be used as an aid during the design process.

Tolerable crack widths, reinforced concrete (ACI 224R-90)	
Exposure condition	**Tolerable crack width in inches**
Dry air or protective membrane	0.016
Humidity, moist air, soil	0.012
Deicing chemicals	0.007
Seawater and seawater spray: wetting and drying	0.006
Water-retaining structures	0.004

Because the table often is incorrectly used (typically by attorneys) to judge cracks in the field, I prefer to label the table as "Average Design Crack Widths." As a designer, I often determine if a crack width is expected to fall within these ACI guidelines. However, observed crack widths shouldn't be compared blindly to the ACI 224 table.

Crack-control equations predict the probable maximum crack width. This usually means that 90% of the crack widths measured in a reinforced concrete member are less than the calculated value. However, isolated cracks can be more than twice the width of the computed 90% value. The coefficient of variation of crack widths in reinforced concrete members is about 40%. Therefore, isolated crack widths greater than those presented in the ACI 224 table shouldn't automatically be a cause for concrete rejection or reduced pay.

The debate on allowable crack widths is usually related to durability concerns. ACI 224 notes that crack width values presented in the table aren't always a reliable indicator of the potential for corrosion and deterioration. Although it is logical to assume that larger cracks create more corrosion, most researchers cannot find a correlation between crack width and corrosion potential.

Remember, also, this discussion relates to reinforced structural members. Therefore, don't use the ACI 224 tolerable crack widths to judge unreinforced or lightly reinforced slabs on grade or unreinforced basement walls. As for residential concrete, anything goes. Some homeowners find a ½-inch-wide crack tolerable. The more typical homeowner, however, will object loudly and at length over a crack of any width that appears in a basement wall or slab on grade.

Chapter 24

Miscellaneous

Q. *We're designing a retaining wall that consists of precast concrete units stacked on each other. What's the coefficient of friction for concrete on concrete? The manufacturer of the units says it's 0.75, but that sounds high to me.*

A. The manufacturer's number is slightly less than the values given by the Portland Cement Association (PCA) and Precast/Prestressed Concrete Institute (PCI). PCI's *Design Handbook* says the concrete-to-concrete friction factor for dry conditions is 0.80. The latest edition of PCA's *Concrete Masonry Handbook*, Appendix A, gives a precast concrete-to-concrete masonry friction coefficient of 0.4 based on a safety factor of two.

Friction factor for concrete on concrete

Q. *I've heard that it's possible to use concrete in making kitchen and bathroom counter tops. Do you know where I can get information on how they're built?*

A. A procedure used by homebuilder Thomas Hughes is described in the August/September 1994 issue of *Fine Homebuilding* (pp. 86-89). You can request a reprint of the article from The Taunton Press, P.O. Box 5506, Newtown, CT 06470 (800-477-8727).

The May 1998 issue of *This Old House* includes an article by Brad Lemley, "The Perfect Countertop?"

How to build concrete counter tops

Q. *What is the R-value of a typical 8-inch-thick, cast-in-place, residential basement wall?*

A. The R-value of an uninsulated, 8-inch-thick basement wall built using normal-weight concrete is 1.35, based on data from the 1993 American Society of Heating, Refrigeration, and Air-Conditioning Engineers' Handbook. By doubling the thickness of the wall to 16 inches, the R-value only increases by 0.50.

R-value of concrete

Q. *Our company does commercial work that often requires us to construct bollards, which are simply 4- or 6-inch-diameter steel pipes filled with concrete. Concrete is rounded at the top so water will run off. The problem is that our workers don't produce a very uniform product. On some bollards the protruding concrete looks like a scoop of ice cream on a cone, and on others the concrete is nearly flat. I know that the bollards' main function is preventing cars from running into storefronts, but the nonuniform appearance is unsightly. Are there any standards for the proper height to mound the concrete? And is there a device that can be used to give the bollards a uniform appearance?*

Standards for bollard construction

A. We checked the most recent edition of The American Institute of Architects' *Architectural Graphic Standards*, and found no mention of a standard height for the curved top of a pipe bollard. We also checked with several finishers, who said they usually just mound up a 4- to 5-inch-slump concrete at the top of the steel pipe and finish it with a small pool trowel. One suggested getting a metal bowl of the required diameter and welding trowel handles on opposite sides, so the concrete could be consolidated and formed by twisting the bowl. If any readers know of a way to produce uniform-looking concrete-filled pipe bollards, please tell us about it.

Reader response: I've found that mounding concrete about 2 or 3 inches above the steel pipe will result in a good-looking finished product. That's because troweling the concrete compacts it. After shaping the concrete with the trowel and letting it set enough to finish, take a piece of plastic film and use it to buff the top and sides of the concrete as if you were shining your shoes. This will fill in any imperfections and leave a nice surface texture. If you like, you can then pass a finishing brush over the surface.
Raymond Furtivo, Pittsburgh

Years ago, I was shown an easy method for putting a uniform top on a concrete bollard. Wearing rubber gloves, mold the top with your hands to the desired shape or height. Next, take a piece of plastic film 16 to 24 inches long and 2 to 4 inches wide (a piece of 6-mil polyethylene works very nicely) and, with one end in each hand, place the film over the fresh concrete and gently work it back and forth and around the top to shape and finish.

As with any finishing operation, you need to get the feel of it and adjust your motions and speed to fit conditions. To get a really nice finish, go over the bollard top several times as the concrete sets. The more passes, the finer the finish. The only time I have trouble with this method is when using a concrete mix with large aggregate. You need to make sure you have some fines over the rock when you do the initial shaping.
Ron Mischnick, Walter Mischnick Contractors & Builders Inc., Alliance, Neb.

We get very good results finishing the tops of concrete bollards. We use a 2- to 3-inch-slump concrete and mound it 6 inches. We then rough-shape the concrete with a magnesium float and finish it with a sheet of 4- or 6-mil plastic. We slide the plastic over the concrete surface in alternating directions.
Herbert W. Katt, president, Katt Construction, Racine, Wis.

To finish concrete bollards, we take a hollow rubber ball, such as a tennis ball, and cut it in half. Next, we twist the half-ball over the concrete in a circular motion to provide a uniform finish. We've built hundreds of bollards using this method.
William Wells, Bulls Gap, Tenn.

Go to the toy store and buy a fairly stiff hollow rubber ball slightly larger than the bollard. Cut in half, the ball makes the perfect tool for shaping the bollard top.
Scott Engler, W.G. Clark Construction Co., Seattle

Effect of overtime on worker productivity

Q. *To get a paving job done on time, we're planning to have our crews work six 10-hour days per week. However, I'm concerned about the cumulative effect of long hours on worker productivity. Is there any data on productivity and overtime specific to the construction industry?*

A. While there is some published data specific to the construction industry, results aren't conclusive. The following information on overtime effects comes from a productivity summary report furnished by consultant James J. Adrian.

Point of No Return

Overtime Schedule, Days (hours per day)	Point of No Return Days	Optimum Stopping Point Days
5(9)	21	14
5(10)	12	8
6(8)	10	7
6(9)	8	5
7(8)	4	3

A study performed for the Construction Industry Institute involved a number of construction projects and crafts that included pipefitters, electricians, formwork carpenters, ironworkers placing rebar, and laborers for concrete placement. This study showed that productivity losses from overtime aren't automatic and that it may be possible for crews to work 60-hour weeks for several weeks without serious negative consequences.

A concept termed "the point of no return" is sometimes used to describe productivity losses due to overtime. The point of no return is the duration of the overtime schedule beyond which the cumulative output is no more than would have been achieved in a 40-hour week. Manufacturing-industry research found points of no return to vary from four to 21 days, as shown in the table above.

A Business Roundtable report suggested points of no return for 50- and 60-hour weekly schedules of seven and nine weeks respectively, significantly higher than the values in the table. And a 1979 study by the American Subcontractors Association, Associated General Contractors, and Associated Specialty Contractors showed greatly differing results. The study concluded that, in general, productivity is greater than normal for the first few weeks of overtime. It also showed that seven weeks into overtime, worker productivity was still equal to the productivity expected for a 40-hour week. The point of no return wasn't reached until the 16th week of overtime. The summary notes that this latter value seems inconsistent with results of other studies.

There is much uncertainty regarding effects of overtime on productivity. Many studies citing detrimental effects are based on limited data and projects. And it's possible that some productivity losses result, not from fatigue or similar causes, but from the contractor's inability to provide materials or information at an accelerated pace.

Load-bearing lawns

Q. *We need more information about the "grass plus concrete" load-bearing lawn, described in* Concrete Construction, *September 1988 (page 863). Who is familiar with the system in Quebec? Who is the maker of the reusable plastic form unit?*

A. The licensee for this proprietary system in the United States works with the reusable former. The company is:

Bomanite Corporation
P.O. Box 599
Madera, California 93639

The original patent holder is a British company:

Grass Concrete International Limited
Walker House
22 Bond Street
Wakefield, Yorkshire WF1 2QP
England

According to company chairman, R.M. Walker, the original system is based on a single-use former. At present there are no licensees in any part of Canada, so the sources above must be contacted.

Making a new bridge capture an old look

Q. *We need to replace two arch bridges in a local park. One is a three-span structure with a 70-foot center span flanked by two 56-foot spans. The smaller bridge is a single 35-foot span. These are both masonry bridges, vintage about 1920, and the park district is concerned that the new bridges maintain some of the aesthetic qualities of the old ones. At the same time, we must find an efficient construction method. What can you suggest?*

Parallel arches built by shotcreting over an inflated form can be back-filled to make a rustic bridge. For enhanced appearance the headwalls may be faced with stone masonry veneer.

A. The City of Grand Rapids, Michigan, faced a similar problem on a somewhat larger structure (*Concrete Construction*, May 1989, page 459). The consultant, Williams & Works, Inc., designed a bridge using precast prestressed I beams with a cast-in-place concrete deck (see drawing). By integrating precast, conventionally reinforced arch-shaped fascia panels into the design, they maintained a resemblance to the old arch bridge that was replaced, and a feel-

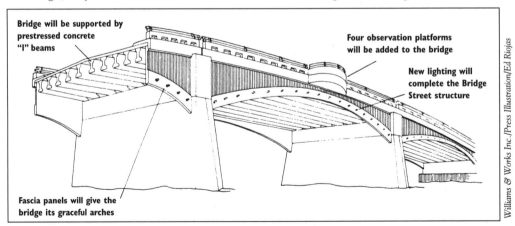

Precast prestressed I beams make up the structure of this bridge. The addition of precast arch-shaped fascia panels makes its appearance more harmonious with older structures in the area. Using standard, readily available bridge beams saved both time and money.

ing of harmony with older structures in the downtown area. Accent lighting which outlines the arches brings the structure to life after dark.

A solution like this gives the owner a wide choice of finishes on the arch face panels, including a ribbed texture that could be achieved with form liners or the texture of aggregate exposed by use of chemical retarders or mechanical tooling. Application of stone masonry veneer as shown on the pedestrian bridges in *Concrete Construction*, January 1990, page 64, would be an appealing but more costly alternative.

For the shorter arch span you might consider an inflated forming method (described in *Concrete Construction*, April 1988, page 385) which permits shotcreting an arch that supports H10 loading.

Q. *We snapped a chalkline in the wrong place on a concrete driveway before sawing joints. We've tried pressure washing and scrubbing, but can't remove the stain. Is there any other way to get rid of the unwanted line?*

A. Two chalk manufacturers print the same information on chalk bottles. "Warning: Red, yellow, and fluorescent chalks are permanent colors. There is no known way to remove them. Blue chalk may also be permanent on porous materials." A lumber-crayon manufacturer said that sun and acid rain might eventually lighten the stain.

Reader response: In a November 1997 *Concrete Construction* Problem Clinic item (p. 928), a contractor wants to know how to remove unwanted chalk lines from a concrete driveway. My theory is "if you can't kill the bull, you best run the other way." Since the errant chalk line can't be removed, I suggest adding more lines to create a design. An architect could sketch an interesting and unique design based on the "less is more" concept, using a few strategically placed lines. Then the chalk lines won't look like a mistake, and the sun will eventually fade them. Also, this fix can be cost-effective. (But you may want to hire someone to place the lines.)
David A. Branch, Henry Tyler Construction Corp., Birmingham, Ala.

Q. *We had some complaints on a recent job about ammonia smells coming from concrete placed in a confined area. What causes this smell?*

A. Ammonium compounds are found in some water-reducing admixtures. Finishers and others have commented about this smell, but we haven't seen information about it being hazardous. As a general rule, check for adequate ventilation on any construction job in a confined space.

Q. *I am using a flat saw equipped with a diamond blade to cut joints in a reinforced-concrete slab. The blade is cutting very slowly, but is showing little wear. I am using plenty of cooling water. What can I do to increase cutting speed?*

A. Diamond blades don't actually cut concrete; they grind away the surface under the blade. The tiny diamond segments are held on the blade by a metal matrix or bond. After several thousand passes through the material being cut, the diamond segments begin to crack and break off. The matrix wears away, releasing the fractured diamonds and exposing new ones. This process continues until the blade is worn out.

Removing chalkline marks

Ammonia smell from fresh concrete

Diamond blade cuts too slowly

An experienced saw operator or diamond blade supplier can look at a blade under a magnifying glass and tell you how the blade is working. If the blade is cutting slowly, its matrix probably is too hard and not wearing quickly enough to expose new diamonds. Try cutting back on the coolant water or switch to a blade with a softer matrix. If changing saws is an option, a saw with more horsepower can increase production.

Another cause of your problem could be too much reinforcing steel in the area being cut. The blade must always cut through concrete to be effective.

Sources for Georgia buggies

Q. *Where can we buy the hand-pushed, two-wheeled buggies used to place concrete? I believe they're called Georgia buggies.*

A. First try your local construction supply house. Most of them carry this type of product. If they don't have Georgia buggies you can contact the manufacturer directly. Check The Aberdeen Group's buyers' guide for a list of buggy manufacturers.

Source for diammonium citrate

Q. *I want to use diammonium citrate to remove discoloration from newly placed concrete. I've called chemical-supply houses and pharmacies, but no one seems to recognize the term. How do I find this material, and how much is it likely to cost?*

A. The most generic term is *ammonium citrate, dibasic.* Try using this chemical name when calling local chemical-supply houses. To get a 1997 price, we called Aldridge Chemical Co., Milwaukee, at 800-558-9160 and found that this supplier calls the product *ammonium hydrogen citrate.* You can buy 2.5 kilograms (about 5.5 pounds) of the prepackaged material from Aldridge for $83.55. If you buy more than 20 pounds in bulk, the price per pound is lower.

Since the recommended solution concentration is 2 pounds of diammonium citrate dissolved in a gallon of water, a gallon of solution costs about $30.

Thawing frozen subgrades

Q. *We do a lot of concrete work in Canada where frozen ground is common. Most specifications prohibit placing concrete on frozen ground. How do concrete contractors in northern climates meet this specification requirement?*

A. The contractor must usually either prevent the ground from freezing or thaw it before placing concrete. The most common method for preventing freezing is to cover the ground with insulation. Insulating blankets and straw have both been successfully used. Another possible insulating method is to cover the ground with a waterproof plastic sheet and construct a pond of water above the subgrade. The water serves as insulation and as a heat sink when it changes to ice.

The amount of water needed to protect the ground from freezing can be calculated. Make sure no water leaks from the pond into the subgrade below. Another solution is to erect an enclosure and provide heat. The enclosure could be a flat tent propped on blocks or it could be large enough to serve as a construction shelter for use during form setting, concrete placement, and curing.

Another way to warm the forms, thaw frozen ground and cure concrete in cold weather is with a portable radiant heat system that pumps a heated biodegradable antifreeze through rubber hoses at low pressure. The hoses can cover up to 6,000 square feet. The system made by Ground Heaters Inc., Spring Lake, Minn., is available in two models, 150,000 and 300,000 BTUs, and is thermostatically controlled.

Also you can build a heated enclosure and wait for the ground to thaw within it. In Canada, frozen ground has been successfully thawed by drilling holes 4 to 5 feet apart and injecting hot water or steam.

You might be able to safely place concrete on frozen ground if the soil isn't frost susceptible. To successfully place concrete on frozen ground you need to prevent the concrete from freezing until it reaches a compressive strength of 500 psi. Adding an extra inch or two of concrete is one option. This could be viewed as a sacrificial layer that may freeze but maintains the heat generated by the rest of the concrete. An alternative is placing the concrete on an inch of rigid insulation board thus protecting it from the frozen subgrade below. Heating the concrete, using Type III cement, or adding an accelerator reduces the time needed to reach the 500-psi strength level. However, placing concrete on frozen ground isn't recommended if the ground isn't thaw stable.

Silts and clayey silts with free access to water aren't thaw stable because they settle after thawing. Also, because of the free water present, these soils are weaker immediately after the thaw. Don't place slabs on grade over frozen soils that aren't thaw stable. Subgrade settlement and decreased bearing capacity may crack and tilt the slabs. Coarse-grained soils such as gravels and sands are typically thaw stable and maintain their strength after freeze-thaw cycles.

If you consider placing concrete on frozen ground, contact a soil testing firm and ask for a thaw-consolidation test on samples of the soil. This test measures the amount of settlement after a soil thaws. Based on the amount of settlement, the soil testing firm can evaluate the advisability of placing concrete over the frozen subgrade.

A simpler test also can be used to help predict thaw settlement. The dry density of frozen and thawed soil is measured. The values are then used in the following equation to predict settlement: Settlement = (1 − frozen density/thawed density) × (height of the frozen layer).

This equation states that if the frozen soil contains more moisture than the thawed soil originally contained, settlement will occur as thawing releases the moisture and the soil shrinks to its original volume.

Do stem walls chemically bond to footings?

Q. *When a stem wall is set on a footer, usually several days after the footer is poured, is there any bonding between the footer and the stem wall that prevents lateral movement of the stem wall? I'm asking this question because some form oil for the stem walls was spilled on the footer surface. The inspector said this reduces bond, even though there was a shear key in the footer.*

A. Chemical bond provides little if any resistance to lateral movement between a stem wall and footing. The resistance is largely mechanical through concrete-to-concrete friction, shear keys, or rebar dowels. The effect of form oil is probably inconsequential.

Differences between wet, moist, and damp subgrades

Q. *On different occasions, I've been told to wet, moisten, or dampen the subgrade prior to concrete placement. What is the difference between wet, moist, and damp as it pertains to the subgrade?*

A. The American Concrete Institute's "Cement and Concrete Terminology" (116R-90) defines each term. Wet is defined as "covered with visible free moisture," damp is a "moderate covering of moisture," and moist is "slightly damp but not quite dry to the touch." Thus wet indicates the highest level of moisture and moist indicates the lowest level.

For subgrade preparation, many engineers and contractors use the three terms interchangeably. However, the term wet should not be used. Most specifiers agree that there must be no free water standing on the subgrade, nor should the subgrade have any muddy or soft spots when the concrete is placed. Therefore, damp is probably the best term to use for subgrade preparation.

Source for flame-cleaning equipment

Q. *We've seen several references to a concrete-cleaning device that removes elastomeric membranes, paints and coatings by flame blasting. Oxygen and acetylene are combined to produce a flame that is passed at a given height and rate over the substrate. This device appears to be suitable for an application we're planning, but we can't locate a manufacturer. Who makes them?*

A. The method you mention, flame blasting, is described in *Guide for Selecting and Specifying Surface Preparation for Sealers, Coatings and Membranes* (Guideline No. 03732), published by the International Concrete Repair Institute. However, the guide doesn't mention a manufacturer for the device you describe. It appears that most contractors who use this method make their own flame-blasting equipment.

How to splice PVC waterstops

Q. *We sometimes have problems with water leakage through construction joints containing polyvinyl-chloride (PVC) waterstops. We believe the leaks are occurring at splices in the waterstop. Is there a good reference on how to splice PVC waterstops so they won't leak?*

A. In July 1995, the U.S. Army Corps of Engineers issued a revised Section 03250 (Expansion Joints, Contraction Joints, and Waterstops) of their *Guide Specification for Military Construction.* Here's what the specification says about splices in section 3.2.3.

"Splices shall be made by certified trained personnel using approved equipment and procedures." For nonmetallic waterstops, "fittings shall be shop-made using a machine specifically designed to mechanically weld the waterstop. A miter guide, proper fixturing (profile dependent), and portable power saw shall be used to miter cut the ends to be joined to ensure good alignment and contact between joined surfaces. The splicing of straight lengths shall be done by squaring the ends to be joined. Continuity of the characteristic features of the cross section of the waterstop (ribs, tabular center axis, protrusions, etc.) shall be maintained across the splice."

For polyvinyl-chloride waterstops, "splices shall be made by heat sealing the adjacent waterstop edges together using a thermoplastic splicing iron utilizing a nonstick surface specifically designed for waterstop welding. The correct temperature shall be used to sufficiently melt without charring the plastic.

The spliced area, when cooled, shall show no signs of separation, holes, or other imperfections when bent by hand in as sharp an angle as possible."

The specification prohibits edge welding and requires centerbulbs to be compressed or closed when welding to noncenterbulb-type waterstops. The specification also lists the following unacceptable waterstop splicing defects:

1. Tensile strength less than 80% of parent section
2. Free lap joints
3. Misalignment of centerbulb, ribs, and end bulbs greater than $\frac{1}{16}$ inch
4. Misalignment that reduces waterstop cross section more than 15%
5. Bond failure at joint deeper than $\frac{1}{16}$ inch or 15% of cross section
6. Misalignment of waterstop splice resulting in misalignment of waterstop in excess of $\frac{1}{2}$ inch in 10 feet

7. Visible porosity in the weld
8. Charred or burnt material
9. Bubbles or inadequate bonding
10. Visible signs of splice separation when cooled splice is bent by hand at sharp angle

Q. *We delivered concrete to a poured wall contractor building residential foundations in a subdivision. He says we shorted him about ¼ yard per 9-yard load.*

I ran a unit weight test on the fresh concrete and divided the total batch weight by the unit weight. The yield came out to be exactly 27 cubic feet per cubic yard delivered, but the contractor still thinks we shorted him.

Is there a standard waste allowance that contractors use when estimating the amount of concrete they need? If so, is the ¼ yard per 9 yards within that allowance?

A. Wasting ¼ yard per 9 yards calculates out to 2.8% waste. The book *Concrete Construction & Estimating* says to include 3% waste for concrete placed on the ground but no waste for concrete placed in forms. However, the book also says: "If an item calls for 31¼ cubic feet of concrete, call it 32, picking up part of a cubic foot on each item. This is how to include waste on the job. But you should still add 3% additional waste for all concrete placed on earth or fill."

In the example quoted, the estimated waste then is 0.75/31.25 = 2.4%.

In the book *Estimating & Project Management for Small Construction Firms*, the normal waste for foundations is given as 5% and for superstructures, 2%. The book also suggests ignoring any openings in walls and slabs less than 15 square feet but deducting from quantity take-off for larger openings.

Some contractors figure their concrete requirements closer than others and might have no more than a couple wheelbarrow loads left over after pouring a 40-cubic-yard foundation. Even at 4 cubic feet per wheelbarrow, that's a waste factor of less than 1%.

References
1. *Concrete Construction & Estimating*, Craftsman Book Co., P.O. Box 6500, Carlsbad, CA 92018.
2. *Estimating & Project Management for Small Construction Firms*, Van Nostrand Reinhold Co., 115 Fifth Ave., New York, NY 10003.

Reader response: Your answer to the question concerning yield completely missed this point. If the contractor in question has measured the wall correctly, and it calls for 36 cubic yards, then four loads of 9 cubic yards should fill that wall.

The supplier cannot be excused for being 1 cubic yard short just because the contractor included a waste allowance. Waste allowance is needed for a myriad of other reasons and is not intended to reward a supplier for delivering short loads.

James Simonetti, Cleveland Cement Contractors Inc., Cleveland, Ohio

Q. *We have to cut a 2x2-foot opening in an existing elevated reinforced-concrete slab. An engineer has approved the location of the opening, but we're trying to decide the best way to cut it. In the past we've drilled core holes at the four corners of an opening, then cut from hole to hole. Coring eliminates overcutting and may reduce stress concentrations at the corners, helping to prevent cracks at the re-entrant corners. Now that concrete chain saws capable of plunge cutting are available, we can make these openings without overcutting or drilling core holes. But is there still an advantage to drilling core holes to minimize the chance of cracking at re-entrant corners?*

A. In a reinforced-concrete slab, cracking at re-entrant corners shouldn't be a major problem since the reinforcing steel will prevent any cracks from widening appreciably. We'd opt for the cutting method that's most economical.

Does concrete expand?

Q. *I attended a seminar at which a speaker said that concrete doesn't expand. I told him I've always read that concrete expands when the temperature rises or when the moisture content increases. The speaker held his ground, however, and said that concrete doesn't expand. Is my literature outdated?*

A. When it first dries, concrete shrinks and undergoes structural alterations that make some of the shrinkage irreversible. Thus, even if it is later resaturated, the initial drying shrinkage isn't fully recovered. Because of this, some concrete-industry people say that once a concrete slab has dried, it never gets larger than its initial volume unless there's an abnormal expansion such as that caused by reactive aggregates. That may have been the point the speaker was trying to make. However, concrete does indeed expand when it gets hot or when the moisture content changes. That's why you need expansion joints in bridges, buildings, and other structures. In exterior concrete, joints widen during cold weather because of cooling contraction and get narrower during hot weather as the concrete expands. If the joints fill with incompressible material during the winter, concrete expansion during the summer can cause pavement blowups.

Understanding truck specifications provides economy

Q. *I am considering purchasing new ready mix trucks. Because weight law changes have taken place in many parts of the country, and EPA emission standards are constantly tightening, how do I know which details to consider when writing specifications for the truck I plan to purchase?*

A. Chassis cost can vary drastically depending on several questions. Before buying a truck, think about questions such as: How much power do I need? How much payload do I need to obtain to make the numbers right? What is the right top speed? What percent of my delivery is paving-curbing?

For example, if 300 hp, 1,150 foot-pounds of torque will do the job, you will save money and weight by avoiding the big block engines. Also, remember the mechanical and legal liabilities involved with excess speed and power.

The added cost of aluminum components is good or bad, depending on how close you are to your payload goal.

Other questions you might ask yourself before specifying a ready mix truck include:
- **The wheelbase:** Do I have the proper wheelbase length to make the necessary internal bridge distance? Will this wheelbase provide the proper cab-to-axle distance for desired weight distribution?
- **The frame:** Do I have the correct amount of after-frame needed for the body supplier? Do I have the crossmembers positioned properly in the after-frame and is air service needed to the end of frame?
- **Cab clearance:** Do I have the proper cab clearance for top rails, rear engine pto, or flywheel pto? Are the air ride cab components out of the way?
- **Air tanks:** Are the air tanks positioned properly, as well as battery and step designs? Make sure the tow hooks won't have to be moved.
- **Pto:** Do I have the proper pto provisions for front or rear engine?
- **Fuel:** How many gallons of fuel will work in our operation so I don't unnecessarily waste payload?

- **Tires:** Will my tire selection handle the GVW or qualify for 500 or 600 pounds per inch of tire width if applicable? Are they rock throwers? Are they set to provide proper wheel cut?
- **Rear axle ratio:** Will the rear axle ratio and tire size fit together with transmission and engine rpm for low and top performance?
- **Service:** Did I arrange for the inservice to be completed before or after the body is mounted? These are just a few things in a constantly changing truck industry you must check on. Get involved with your specification and it will save you time and money.

Q. *What practices can be used when placing concrete to help make buildings more resistant to radon?*

Constructing radon-resistant buildings

A. As you probably know, solid concrete is an excellent material for use in constructing radon-resistant buildings. If cracks are properly sealed, concrete is an effective barrier to the penetration of soil gases. In concrete construction, according to information compiled by the NRMCA, the critical factor is to eliminate all entry routes through which gases can flow from the soil into a building. Here are some guidelines:

1. Design to minimize utility openings; sump pumps should be sealed over and vented outdoors.
2. Minimize random cracking by using control and isolation joints in walls and floors. If done properly, any cracks will occur at the planned joints and can be sealed easily.
3. For slab-on-grade homes in warm climates, placing foundations and slabs as a single monolithic unit is an effective way to minimize radon entry.
4. Use materials and practices, such as larger aggregate sizes and proper water-cement ratios, that will minimize concrete shrinkage and cracking
5. When using polyethylene film beneath a slab, place a layer of sand over the polyethylene.
6. Remove grade stakes after striking off the slab.
7. Construct joints to facilitate caulking.
8. Cure the concrete adequately.
9. Caulk and seal all joints and openings in walls and floors.

Q. *During the winter, we have trouble with our pneumatic vibrators and tools freezing up because of the cold temperatures. What can we do to prevent this problem?*

Tools freezing in cold weather

A. In cold weather, pneumatic vibrators and tools are subject to freezing, particularly when used in air systems with excessive moisture in the lines. Thawing can be accomplished by introducing any one of several commercially available deicing agents into the air line close to the tool.

Care should be taken to follow the manufacturers' instructions and warning as some deicing agents can be poisonous. Make sure these agents are compatible with the required lubricating oil, airhose and tool, or vibrator. Tanner Systems Inc., P.O. Box 59, Sauk Rapids, MN 56379 (320-252-6454), is the maker of one such deicing agent, but there are others available.

Q. *Our aggregate supplier is thinking of switching weight measurements from metric tons to megagrams. What is a megagram?*

Megagrams or metric tons?

A. A megagram, abbreviated as Mg., is an alternative term for a metric ton, based upon the mass' weight of 1,000,000 grams. Since both terms describe the same mass measurement, many metrication experts are urging adoption of megagram in place of metric ton (or tonne) to avoid confusion.

As many states continue to switch to metrics, the term "ton" may be interpreted as one of several measurement units. For instance, the customary inch-pound measure for a large mass is the short ton equaling 2,000 pounds. This is sometimes confused with the long ton (2,240 pounds), metric ton and tonne (both of which weigh in at 2,240.6 pounds) and register ton, a shipping term that's actually a measurement of volume, not weight.

For more information on the megagram and other current construction metrication issues, contact the Construction Metrication Council of the National Institute of Building Sciences at 202-289-7800, or view its Web site at www.nibs.org.

Controlling drilling slurry in occupied areas

Q. *A building owner wants us to drill a large number of holes in concrete walls between offices so they can network computers. How can we control slurry spray when drilling these holes so workers can remain in their offices?*

A. Skilled operators can drill wall cores in occupied areas with little disruption to the occupants and building contents. One coring specialist says that spray of drilling slurry usually occurs when the core barrel starts to cut the concrete and when the core is removed. He uses minimal cooling water until the cut is established. To catch water flowing out the hole, he uses duct tape to make a combination flange and funnel and diverts the cutting slurry into a bucket.

At the end of the cut a moist ring usually appears at the core location just before the barrel cuts through. If possible, have someone watch for this sign and notify the driller to stop. The driller can then tap on the core from the intact side and break it loose. If the driller must work alone, he or she should cut to a depth less than the wall thickness, then break out the core.

Waterproofing vs. dampproofing

Q. *When reading various product literature, I often run across the words "waterproofing" and "dampproofing." They seem to be used interchangeably. Is there a difference? If so, which products are waterproofing products and which are dampproofing? For what applications are each needed?*

A. According to ACI 515.1, waterproofing is a treatment of a surface or structure to prevent the passage of water under hydrostatic pressure. Dampproofing is a treatment of a surface or structure to resist the passage of water in the absence of hydrostatic pressure. As a guideline, waterproofing is often needed for below-grade applications, whereas dampproofing is often satisfactory for above-grade applications.

Traditionally, waterproofing barriers consist of multiple layers of sheet-applied materials impregnated with asphaltic, bituminous, or coal tar material. Today, a number of other waterproofing barriers can be selected, which include cold-applied systems such as elastomeric membrane barriers, cementitious membranes, modified bituminous materials, bentonite-based materials, and a number of proprietary types of waterproofing barriers. The number of coats or plies, thickness, and types of materials will vary with job conditions.

Dampproofing materials include most sealers, paints and coatings, and water-repellent admixtures and cements. However, their effectiveness varies depending on the material and the application thickness or dosage rate.

There is, of course, a lot of gray area. When determining a system's ability to repel water, it is probably best to view the products on a continuous scale rather than attempting to define them as either waterproofing or dampproofing materials.

Q. *What are the new North American Industrial Classification System (NAICS) codes for aggregates and concrete construction businesses?*

Aggregate, concrete construction get new codes

A. The U.S. Bureau of the Census recently published a new coding system for business activity that replaces the 1987 four-digit SIC code system. (See table).

Under the old system, aggregate businesses were coded with 1422. The NAICS code splits aggregate producers into two classifications: 212312 for producers of crushed and broken limestone, mining and quarrying and 212321 for producers of sand and gravel.

Highway and street construction contractors (except for elevated highways) formerly classified by the SIC code of 1611 are now classified under the NAICS code of 234111.

1987 SIC code	NAICS code
1741 Masonry, stone setting and other stone work	23541 Masonry and stone contractors
1771 Concrete work	
	23542 Drywall, plastering, acoustical and insulation contractors (includes stucco)
	23571 Concrete contractors
3271 Concrete block and brick	327331 Concrete block and brick manufacturing
3272 Concrete products, except block and brick	327332 Concrete pipe manufacturing
	32739 Other concrete product manufacturing
	327999 All other miscellaneous nonmetallic mineral product manufacturing
3273 Ready-mixed concrete	32732 Ready-mixed concrete manufacturing

Q. *We have a number of tilt-up panels, 8 inches thick, lying in place on a 6-inch floor slab. We can't get them free of the base slab so that we can start tilting. It seems that someone forgot to put bond breaker on the foundation slab. We were unsuccessful in wedging the slabs free. What else can be done?*

Bond breaker forgotten on tilt-up job?

A. Wedging, which you tried, is the usual remedy, and it works well on captive panels that are not too strongly bonded. Steel wedges are driven in at the top edge of the panel and at insert lines while attempting to lift the panel and slowly peel it off. The initial pull should be more than the lift weight but not enough to jerk the panel up. It is good practice to pause briefly after the initial pull on the panel. A minute or so is sometimes needed before the panel can overcome bond with the casting surface.

The initial release of a tilt-up panel from its casting bed is an especially sensitive stage. Here the operation was successful, and workers are taking control of braces that were attached before the lift started.

Small panels that stick have been successfully lifted by jacking parallel to the floor slab. This method rarely works with large panels because it is difficult to find an object large enough to remain stable at the base of the jack.

A third possibility that has worked on occasion with both large and small panels is freezing with dry ice. Fist-size pieces of dry ice are spread uniformly over the panel at the rate of about ⅓ pound per square foot of panel. A 6-mil layer of polyethylene film is then placed over the dry ice, forming a tent. Contraction of the concrete and the reinforcing steel sets up a shear plane between the casting slab and the panel. The time required varies, depending on the ambient temperature, the amount of dry ice, and the severity of the sticking problem.

If there is a crane on the job overnight, the panel may be stacked with dry ice, with crane lines hooked to the panel and a slight amount of tension applied. Reportedly, the panel is usually free by the following day.

We queried Peter Courtois of Dayton Superior Corp., a past-president of the Tilt-Up Concrete Association. He confirmed wedging, jacking, and freezing as the available alternatives for releasing problem panels. However, he expressed the view that on some large jobs the economical alternative might be simply to fill in any openings in the tilt-up panels and raise the grade of the floor slab.

Considering the horror of a tilt-up panel not separating from the floor slab, it is rare that the bond breaker is actually forgotten. However, in hot, windy weather some added precautions may be required. If the base slab surface begins to dry before the bond breaker-curing agent is put down, the material can be drawn into the concrete pores instead of remaining on top. Under such conditions it is a good idea to use a fog-mist spray on the floor slab to fill the concrete pores before the curing and bond breaking agent is applied. It also could help to apply a second coat of bond breaker at right angles to the first, before any reinforcing steel for the tilt-up panel is in place.

Extra-high-strength block

Q. *We recently came across a specification calling for extra-high-strength concrete block. We have not heard of this type of block. What is the difference between high-strength concrete block and extra-high-strength concrete block?*

A. According to Steve Kosmatka, manager of research and development for the Portland Cement Association, high-strength block can be manufactured with most aggregates or combinations of aggregates by all block producers. Extra-high-strength block, on the other hand, is usually limited to applications where it is required to limit wall thickness in buildings over 10 stories high.

High-strength and extra-high-strength block are not yet defined by a national specification, but they have been considered to fall into the strength classes shown in the table.

Concrete block strengths		
Type of block	Gross area strength, psi	Net area strength (53% solid units), psi
Regular strength	1060	2000
High-strength	1860	3500
Extra high strength	2650	5000
Source: Portland Cement Association		

Q. *What is the difference between cast stone and precast concrete? Or is there a difference?*

Cast stone differs from precast concrete

A. The American Concrete Institute's glossary, "Cement and Concrete Terminology" (ACI 116R-85) defines cast stone as concrete or mortar cast into blocks or small slabs in special molds so as to resemble natural buildingstone. In the same glossary, precast concrete is defined as concrete cast elsewhere than its final position.

But W.N. Russell III, executive director of the Cast Stone Institute, says those definitions aren't complete enough. He says cast stone is a highly refined architectural precast concrete building stone manufactured to simulate natural cut building stone. The most significant dif-

W.N. Russell and Company

Careful manufacturing methods for cast stone permit perfect matching of repairs when older buildings are restored. Here, cast stone replaced missing pink granite cornice and terra cotta balcony details.

ference between architectural precast concrete and cast stone is that cast stone isn't permitted to contain bugholes or air voids and must have a fine-grained texture. The texture is normally achieved by acid etching. Cast stone is normally used as a masonry product and is included in the masonry section (04435) of standard specifications. Architectural concrete is included in the precast concrete section (03450) in specifications.

Store bull floats out of the sun to prevent warping

Q. *What causes my bull floats to warp within a couple of months after purchasing? I buy magnesium floats with ribs to protect them from warping. I store the floats in the bed of my truck, leaning against the cab.*

A. Magnesium bull floats are susceptible to warping if they are not cared for properly. Always store a bull float in a flat position, out of the sun. If the float lies in the sun and one side heats up faster than the other, permanent deformation can occur.

Volume of grout

Q. *We're pumping a grout that's made with 1 cubic foot of cement and 5 gallons of water. There's no sand or gravel in the grout. What will the volume be?*

A. You can calculate the volume if you know the specific gravity of the cement. Specific gravity is the weight of the cement divided by the weight of an equal volume of water. If you're using portland cement, 1 cubic foot loose volume weights 94 pounds and has a specific gravity of 3.15. Dividing the weight, 94 pounds, by the specific gravity times the unit weight of water (62.4 pounds per cubic foot) gives a volume of cement of 0.478 cubic feet.

A gallon of water weighs 8.33 pounds. Dividing the weight of 5 gallons of water (8.33x5) by 62.4 pounds per cubic foot (pcf) gives 0.667 cubic foot of water. The volume of the grout is the sum of the volume of cement and water or 1.145 cubic feet. This assumes that there's no air entrapped in the paste. If air is entrapped, the volume will be greater.

You can also calculate volume without knowing cement specific gravity if you first measure the unit weight of the grout. To do this, fill a quart jar with the grout, using a small glass plate over the jar mouth to ensure that the jar is exactly full. Weigh the jar, grout, and glass plate on a postal package scale or other scale that can weight to 0.01 pound. Then clean out the jar and weigh the jar and plate alone. Calculate the weight of the grout and divide it by the volume of the jar to get the unit weight. Then divide the weight of a grout batch by the grout unit weight to find the batch volume. Here's an example calculation:

Weight of grout, glass jar, and plate	5.01 pounds
Weight of glass jar and plate	1.15 pounds
Weight of grout	3.86 pounds
Volume of quart jar	0.033 cubic foot
Unit weight of grout (3.86/0.033)	117 pcf

If you use 41.7 pounds of water (5 gallons) and 94 pounds of cement per batch, the grout volume will be:

(41.7 + 94)/117 = 1.16 cu. ft.

Quart canning jars may not hold exactly a quart. If you want a more precise test for unit weight, calibrate the jar first by weighing it filled with water. Divide the weight of the water in pounds by 62.4 pcf to calculate the exact volume of the jar.

Q. *What is a* crandalled surface *as specified in* Specifications for Structural Concrete, ACI 301-96, section 6.3.7.3.c, "Tooled Surfaces"? *The term is also used in the U.S. Army Corps of Engineers' Guide Specification but isn't defined in either of these documents.*

Definition of a crandalled surface

A. Bryant Mather of the Corps of Engineers found the following definitions for *crandall* in *A Dictionary of Mining, Mineral, and Related Terms*, published in 1968 by the U.S. Bureau of Mines:
- "A stonecutter's hammer for dressing ashlar [a block of stone, as brought from the quarry]. Its head is made up of pointed steel bars of square section wedged in a slot at the end of the iron handle.
- "To dress stone with a crandall."

It appears that concrete as well as stone can be crandalled and, from the description, a crandalled surface seems to be similar to a bush-hammered surface.

Q. *What do ACI and ASTM standards say about adding water to ready-mixed concrete at the jobsite? Is it permitted? If so, when and how should it be added?*

Adding water to concrete at the jobsite

A. ASTM's "Standard Specification for Ready-Mixed Concrete" (ASTM C 94-90) permits one addition of water under some circumstances. Section 11.7 says:

"When a truck mixer or agitator is approved for mixing or delivery of concrete, no water from the truck water system or elsewhere shall be added after the initial introduction of mixing water for the batch except when on arrival at the jobsite the slump of the concrete is less than that specified. Such additional water to bring the slump within required limits shall be injected into the mixer under such pressure and direction of flow that the requirements for uniformity in Annex A1 are met. The drum or blades shall be turned an additional 30 revolutions or more if necessary, at mixing speed, until the uniformity of concrete is within these limits. Water shall not be added to the batch at any later time."

Dick Gaynor of the National Ready Mixed Concrete Association says the words "on arrival" are important. If slump is low, the water addition should be made when the truck arrives at the jobsite—not after it has waited for some time. The water may be added at this time in one or more increments as the specified slump is approached. Once a slump within specification limits is reached, no further water additions are permitted.

ACI's "Specifications for Structural Concrete for Buildings" (ACI 301-89) also address this issue. Section 7.5.2 says: "When concrete arrives at the project with slump below that suitable for placing, as indicated by the specifications, water may be added only if neither the maximum permissible water-cement ratio nor the maximum slump is exceeded. The water shall be incorporated by additional mixing equal to at least half of the total mixing required. An addition of water above that permitted by the limitation on water-cement ratio shall be accompanied by a quantity of cement sufficient to maintain the proper water-cement ratio. Such addition shall be acceptable to the architect/engineer or his representative."

Editor's note: The *Concrete Construction* June 1991 problem clinic answered a question about adding water to concrete at the jobsite. We quoted from ACI 301-89, which says water can be added only if the maximum permissible water-cement ratio isn't exceeded. Dick Elstner wrote from Hawaii to ask how an inspector can know whether the maximum permissible water-cement ratio is exceeded. Because of free moisture on aggregate surfaces, it's difficult if not impossible to know precisely how much mixing water is in the concrete. The batch man corrects for aggregate moisture but then he or the truck driver typically final-adjusts the mix by adding water until the design slump is reached. So how is the inspector to know how much mixing water is present (and thus know the water-cement ratio) of the delivered concrete?

The only way the inspector will know is if all water additions are recorded and the inspector has a record of those additions. Surface moisture contributed by the aggregate can be calculated if the aggregate surface moisture content is measured. This amount of water can be added to the amount batched at the plant, the amount of wash water left in the truck from the previous batch, and the amount of water added by the batch man or the driver.

An accurate estimate of mixing water content requires an accurate determination of aggregate moisture content. It also requires accurate measurement of other water added to the batch. The system is admittedly less than perfect. But until it's possible to quickly measure water-cement ratio of fresh concrete at the jobsite, there aren't many other alternatives when a maximum permissible water-cement ratio is specified.

The more current ACI 301-92 is worded slightly differently. Section 4.3.2.1 says: "When concrete arrives at the point of delivery with a slump below that which will result in the specified slump at the point of placement and is unsuitable for placing at that slump, the slump may be adjusted to the required value by adding water up to the amount allowed in the accepted mixture proportions unless otherwise specified by the Architect/Engineer. Addition of water shall be in accordance with ASTM C94. Do not exceed the specified water-cementitious material ratio or slump. Do not add water to concrete delivered in equipment not acceptable for mixing.

"After plasticizing or high-range water-reducing admixtures are added to the concrete at the site to achieve flowable concrete, do not add water to the concrete.

"Measure slump and air content of air-entrained concrete, after slump adjustment, to verify compliance with specified requirements.

What is green concrete?

Q. *The term* green concrete *is sometimes used in specifications, but it seems as if everyone defines it differently. What does the term mean?*

A. The American Concrete Institute (ACI) defines green concrete as concrete that has set but not appreciably hardened (Ref. 1). Saw manufacturers and sawing contractors refer to green concrete as concrete that has set but hasn't fully cured (Ref. 2). Both definitions are imprecise because the terms *appreciably hardened* and *fully cured* aren't defined.

Hardening refers to strength gain after final set has occurred. Thus, the ACI definition seems to imply that green concrete has reached final set but hasn't gained much strength. The saw manufacturers' definition implies that some hardening or strength gain has occurred because green sawing should begin when concrete won't spall or ravel at the cut edges. Green concrete can also mean freshly placed (presumably plastic) concrete (Ref. 3). However, the term is usually applied to concrete that has lost its plasticity.

If the term is used but not defined in specifications, you should ask the specifier for a definition. Unfortunately, it probably won't be any more precise than the three definitions described above.

References
1. ACI Committee 116, *Cement and Concrete Terminology*, ACI 116R-90, American Concrete Institute, Farmington Hills, Mich., 1990, p. 16.
2. *Target Catalog*, Target Products, Kansas City, Mo., 1994, p. 95.
3. J. Stewart Stein, *Construction Glossary*, 2nd ed., John Wiley & Sons Inc., 1993, p. 112.

Definition of no-slump concrete

Q. *We just ran across a requirement for no-slump concrete. What concretes are considered to be no-slump concretes?*

A. According to ACI's "Cement and Concrete Terminology" (ACI 116R-90), no-slump concrete is freshly mixed concrete exhibiting a slump of less than ¼ inch. Zero-slump concrete is defined in the same document, as concrete of stiff or extremely dry consistency showing no measurable slump after removal of the slump cone. And negative-slump concrete is concrete of a consistency such that it not only has zero slump but still has zero slump after additional water is added.

Q. *I have a residential-concrete specification that calls for a 1:2:4 concrete mix. What does that mean?*

What's a 1:2:4 mix?

A. This is a little-used method for showing the amount of cement, sand, and coarse aggregate in the concrete. The specification should also say if those proportions are by weight or volume.

If the proportions are based on weight, use 4 pounds of coarse aggregate and 2 pounds of sand for every pound of cement. To make the concrete, though, you still need to know the required water-cement ratio.

If the proportions are based on volume, use four shovelfuls of coarse aggregate and two shovelfuls of sand for every shovelful of cement. Batching by volume isn't very accurate because a shovelful of dry sand will put more sand in the mix than a shovelful of wet sand.

Q. *We repeatedly hear that strength of concrete depends almost entirely on the water-cement ratio. Yet I cannot understand why two mixes, one containing four bags of portland cement per cubic yard and the other containing six bags per cubic yard, could have the same strength. The second has 50% more cement than the first. Isn't the second mix really stronger at the same water-cement ratio?*

What makes concrete strong?

A. The two mixes will be very close to the same strength. Suppose that you make two mixes of cement and water only. One has four bags of cement and the other six. Both contain, let us say, half as much water as cement by weight. When they have set and cured, they will both have the same strength, though one will occupy a larger volume than the other.

Now, mix two more batches exactly like the first ones. Into each one add enough saturated surface-dry aggregate to make a total volume of one cubic yard. Obviously, one mix will take a little more aggregate than the other. Now, when they have cured, the weakest link in both masses will be the cement-water portion—the cement paste. The cement paste will determine the strength. But we've noted before that both pastes have the same strength; consequently both concretes will have the same strength.

There are, of course, some slight differences in strength of the two mixes, because there is less aggregate surface in one mix than the other, but the difference is minor.

There is one qualification to the above discussion: The explanation does not apply to concretes in which the aggregate is weaker than the cement paste. This is likely to be true, for example, of concretes made with perlite, vermiculite, or polystyrene aggregate.

Q. *I am always amazed when I read articles concerning the compressive strength of concrete in your publication and others. I've seen recommendations of 3000- to 4000-psi, air-entrained concrete for residential slabs on grade. However, here in Southern California most of our flatwork calls for 2000 psi with no air entrainment. Is this due to the difference in the weather conditions between Southern California and the rest of the country or is Southern California behind in good concreting practices and technology?*

Differences in specified strength

A. Minimum compressive strength is often specified for concrete. It is easily measurable and relates to other important properties such as flexural strength and durability. American Concrete Institute Committee Report 332, "Guide to Residential Cast-in-Place Concrete Construction," offers guidelines for selecting concrete strength. To select a compressive strength you must first choose a region based on the climate and also know the type or location of the construction.

The climate in Southern California is classified as mild. For a driveway constructed in a mild climate, the guidelines recommend 2500-psi concrete with air entrainment to improve workability and cohesiveness. However, where local experience shows a history of satisfactory performance and where local codes permit, 2000-psi concrete may be used.

For driveways constructed in a severe climate, such as the Midwest or the Northeast, it is recommended that a compressive strength of 3500 psi be used. It has been shown, however, that the use of 4000-psi concrete in areas subject to deicer salts gives superior durability.

Although higher compressive strength concrete is often specified to meet durability requirements in areas of severe climates, in your part of the country, where a freeze is a rare occurrence, lower-compressive-strength concrete is probably adequate.

Index

glass-fiber reinforced concrete
    fibers used, 164
green concrete, 238
grinding high spots on floor, 98, 108
grooving
    effect on floor strength, 74
grout
    volume calculation, 236
grout fluidifiers
    for preplaced-aggregate concrete, 4

H

half cell device
    to detect corrosion, 177
hard hat
    replacement frequency, 181
heat of hydration
    effect on form liner, 86-87
heaving
    basement floor slab, 55-56

I

ice skating rink
    durable concrete for, 64
inflatable forms
    for arch bridge, 224-225
    source, 90
infrared thermography
    for detecting delaminations, 198-199
internal vibrators
    coated head prevents damage to epoxy-
        coated rebar, 151
isolation joint
    cracks at, 45

J-K

joints
    asphalt-concrete joint detail, 16
    construction joints should line up, 17-18
    depth of cut for early-cut saws, 19-20
    effect of cold weather on joint sawing,
        16-17
    expansion joint spacing, 18-19
    floor joint repair, 23
    forming isolation joints, 20
    grooved cause snowplow damage?, 15
    in parking-deck slabs, 20-21
    pinwheel, 21
    repair of sidewalk expansion joints, 21-22
    sawcut timing, 15-16

    sealant failure, 22
    shape factor for filled or sealed joints, 22
    spall repairs for floors, 24-25
keyed floor joints
    repair, 23
keyway
    with continuous welded-wire fabric,
        177-178
    for 4-inch-thick slab?, 176

L

latex-modified concrete
    bonded to magnesium-phosphate-cement
        mortar, 135-136
leakage
    between basement floor and wall,
        108-109
    caused by cold joints, 22
lightweight concrete
    finishing problems, 81
lignite
    avoiding surface defects, 3
load-bearing lawns, 223-224

M

magnets
    for formwork blockouts, 88-89
mat test
    for water-vapor emission, 195-196
megagram
    definition, 231-232
mix design
    for exposed aggregate concrete, 9-10
mix proportions
    weight or volume ratios, 239
muriatic acid
    causes driveway deterioration, 187

N-O

nailable concrete, 126
no-slump concrete, 238-239
OSHA
    complaints trigger inspections?, 181-182
overlay
    acid resistance, 73
    bond failure causes, 144
    over cellular concrete, 133
    debonding, 137
    epoxy, effect of UV radiation, 136

Y-Z